T0198011

# TABLE OF CONTENTS

*A family portrait of August 2009 with wife Shen-Ling (second to left), daughters Jennifer (right) and Cynthia (left)*

*A celebration of Google China's new site in 2006*

# Foreword
# Why Don't We Agree to Disagree?

On January 12, 2010 at 7:10 a.m., my phone rang. It was from David Barboza of the New York Times. I picked up the phone, and David asked, "Kai-Fu, I want to talk to you about Google pulling out of China."

I respected David a great deal – he had written several stories about my work at Google, always with exceptional insight and in lightening time. But this was one interview I could not do.

After politely declining his interview, my phone rang again – the Wall Street Journal, Reuters, AP, the Chinese portals, Chinese Dailies…. By the time I returned home that night, my cell phone had run out of battery from ringing all day.

My energy was as drained as the battery in my phone. It was exhausting to see my hard work in four years crumble in one day. My four-year tenure as President of Google China was the most exhilarating experience – hiring 700 brilliant and dedicated people and rallying them to compete against the 7000-person Baidu, China's largest search engine; developing a Silicon Valley culture yet localizing it to fit in China; bringing Google China's market share from 10% to 35% while Yahoo, eBay, AOL and other Internet giants gave up their China ventures; reaching a 2010 revenue level reported to be $600 million; creating great products from the most accurate web search to the most popular map and mobile products that delighted Chinese users.

I said to myself, "If Google pulls out, the products will become inaccessible to most users, and the world's number-one brand Google may become unrecognizable in the world's most populous country."

I asked myself, "What would happen to all the hard work that my team and I put in over those four years? Would things have been different had I not resigned from Google on September 4, 2009?"

That night and every night that week, it was difficult for me to go to sleep – I continued to think about what happened, what made it happen, and what might be next…

More than a year has passed since then. As of October 2011, Google's market share in China has dropped to single-digit. Its services are harder and harder for Chinese users to access.

Looking back, I realize the Google China drama was the perfect manifestation of the never-ending China-America chasm. These two great countries and their people are forever trying to understand each other but end up succumbing to stereotypes; hearing the other's words but not comprehending the meaning; endorsing each other in words but undermining each other in deeds; demanding the other to accommodate and empathize but remaining intransigent itself. In my view, these phenomena all boil down to a lack of understanding.

I remember when I took a prominent lawyer to visit the Forbidden Palace in Beijing. After several hours touring the exquisite grounds, she could no longer hold back her question, and had to ask me, "Kai-Fu, please tell me, where is the emperor?" If a highly educated lawyer does not know that China's monarchy ended in 1911 and the post-1979 China is a socialist state practicing capitalist economy, what hope do we have that two countries will know each other?

I remember when I started Microsoft Research China, and visited the dean of Engineering at the famous Tsinghua University. I asked him if he could send students to be our interns. But the word "intern" made no sense to him, as there had never been any companies trying to hire interns from his esteemed school. After much explanation, something clicked, and he said, "Oh, I get it. Interns. Like Monica Lewinski." If "intern" becomes a symbol of promiscuity, then how can the Chinese truly understand the way the American R&D system became the world's best?

I remember while heading up Microsoft Research China, being swamped by angry Chinese reporters demanding an explanation why Bill Gates called Chinese people thieves. I later learned that Gates had told *Fortune*, "As long as they [Chinese people] are going to steal [software], we want them to steal ours. They'll get sort of addicted, and then we'll somehow figure out how to collect sometime in the next decade." As the world's richest person, Gates usually speaks his mind, and in this case he was angry about piracy. But neither he nor his PR managers understood that what he said not only made Microsoft a scheming company with a conspiracy to dominate China and stifle local competition, but worse, he dishonored and insulted the entire people. Honor (also known as "face") is more important to the Chinese than virtually anything.

So this is why my job as the China executive for American companies was difficult. It was like the job of a diplomat when neither country understood the other, or the job of a marriage counselor trying to help a hopelessly stubborn couple get back together, or the job of a translator between people from two planets.

Over the years, I have learned that if each country could understand the other's history, culture, and viewpoint, and accept that there are some issues that the two countries will "agree to disagree", there would be tremendous progress. I have come to really like the wise Chinese proverb "*yi zhong qiu tong*," which means seeking common ground while accepting differences. This is precisely the mindset that both countries need.

I've done everything I could to help prominent American companies understand China. Now I'm also helping promising companies of China understand America, the world, and China's responsibilities as it rises in this world. I believe China and America will respect each other more if both nations see what I have seen.

Through my personal stories, I hope the Chinese will realize that when Americans appear to be self-righteous, not to assume it is because Americans are aggressive bullies, but consider that it might be because American's desire to help and share their formula for success. American behaviors are shaped by America's rapid rise to prosperity, and a deep sense of righteousness.

Through my personal stories, I hope that Americans will recognize that when the Chinese appear to be autocratic, not to assume it is because Chinese are power-hungry thugs, but consider that it might be from their desire to lengthen stability and pave the ground for a better tomorrow. Chinese behaviors are shaped by China's older glorious history that makes them proud, and also by newer traumatic history that makes them cautious.

So this is why I have decided to write this book about my life and my stories. It is not because my life is so spectacular, but because I hope my stories can shed some light on China and America.

I hope my stories will help the Chinese and Americans better understand each other.

I hope my stories will help the next Google decide to enter China, and the next Chinese Google to enter America.

I hope my stories will help both great countries to seek common ground while accepting differences.

*Speaking as president of Google China at a company event*

# CHAPTER 1

# Following My Heart

Around 8 a.m. on Aug. 5, 2009, the United Airlines flight 888 began landing. It was soon going to arrive at San Francisco International Airport, where I had passed through countless times.

I looked out the window. In front of my eyes was the San Francisco Bay in grayish blue, across which Golden Gate Bridge's red paint softly glistened in the misty morning sun. The scenery seemed the same as every time I had seen it before.

It was hard to believe almost 37 years had passed since my first visit to San Francisco as a boy, and it had been 19 years since my first trip to Silicon Valley for a job interview with Apple. Looking back at those bygone years, I realized my life had been full of changes because I had made unconventional choices.

In 1990, I made the first career change of my life. I gave up a tenure-track assistant professor's position at Carnegie Mellon University to develop new products for Apple. From then on, I made significant breakthroughs in technology, but faced unpredictable market conditions at the same time. I went through ups and downs. However, even in the worst hardships, I never regretted putting myself through ruthless challenges of the corporate world.

My decision to enter the field of high tech business was based on the lifelong motto I had acquired from a philosophy professor in college. Whenever facing a choice, I would always recall his clear voice saying, "Imagine two worlds, one with you and one without you. What's the difference between the two worlds? Maximize that difference. That's the meaning of your life."

Had I stayed in academia, my life would have been much easier, but I would have kept writing papers that no one would read. In contrast, what I have done in product development has influenced the world.

When iTune and QuickTime became hot commodities, I was no longer working for Apple. But I felt a sense of achievement about them as they came out of Apple's Interactive Multimedia Department, which I had created. It didn't matter to me that I wasn't there to share Apple's success in multimedia.

Pioneers don't have to reap in the field they have discovered. A natural born pioneer would rather leave for uncharted territories to make new discoveries.

By summer 2009, I had served as a pioneer for Apple, SGI, Microsoft and Google. I had founded Microsoft Research Asia and Google China. At this point, as Google China prospered, wouldn't it be time for me to move on, to once again explore the unknown and build another venture from the ground up?

I had just completed my four-year term with Google China. In those four years, Google China provided the most accurate and timely Chinese language search, which helped increase Google's market share in China from 16.1% to 31%. Google Maps, Google Mobile Maps, Google Mobile Search and Google Translate all became the most used software products in China. Most notably, Google Music created the world's first advertising-funded legal music download.

It always made me smile when I walked into a coffee shop in China and saw young people downloading songs from Google Music, getting directions from Google Maps, or looking for information through Google's universal search. I knew Google China's business had truly taken off. Our hardest times were over.

I appreciated the more than 700 employees of Google China for all they had done in the past four years. It was their brilliance and persistence that kept turning adversity into opportunity. To me, they were not just subordinates or colleagues. They were all my friends.

They were now waiting for me to continue leading them in the next four years. In the meantime, an email about my contract renewal awaited my reply. The amount of stock in the renewal offer exceeded my expectations. The largest Internet company in the world was expecting me to bring its latest products, namely Android, Chrome and Google Wave, into the China market.

I didn't reply to the beckoning email because I was going to respond to the generous offer in person. As I stepped toward my rental car in the parking lot of San Francisco International Airport, I asked myself, "Are you ready to give Google the answer?"

A cool wind from the San Francisco Bay brushed by me, refreshing my head and my mind. I quietly but firmly told myself, "Yes, I'm ready."

I opened the door on the driver's side of my rental car, sat in, and entered "1600 Amphitheater Parkway, Mountain View" into the car's GPS, which immediately indicated the trip would take 45 minutes.

I started the engine.

## A Deliberate Decision

South of San Francisco International Airport, along Highway 101 were hills and fields bathed in splendid sunshine. The scenery was as serene as my heart.

I drove slowly, with an intent to reassure myself that I wasn't rushing into the decision. I had given it the most thorough consideration.

There was not much traffic on the road. The smooth ride took me through the peninsula of the San Francisco Bay Area. Then I passed by Palo Alto, the northernmost city of Silicon Valley and an affluent town surrounding Stanford University. Mountain View was next. I took the Amphitheater Parkway exit, and arrived at the Google headquarters within a few minutes.

I looked at the giant dinosaur skeleton standing in the central courtyard surrounded by four purple highrise buildings. Why did Google put a dinosaur skeleton there? I never asked. Somehow it seemed self-explanatory to me. The dinosaur skeleton could symbolize a scientific puzzle waiting to be solved, a scientist's never-ending pursuit of knowledge, a science fiction fan's wild imagination, or a simply a childhood dream.

Most Googlers probably used to be inquisitive children who couldn't stop staring at a dinosaur skeleton in a science museum. Many of them still display as much curiosity and energy as children. Working with them brought out the naughty boy in me. I joked with them and laughed with them. I had more fun at Google than at any other workplaces.

I will never forget my first day at Google. A Google chef pushed a cart carrying a five-tier cake into my welcome party, where hundreds of Googlers were enthusiastically celebrating my new start. I felt I belonged here.

After establishing Google China, I often flew back to the headquarters to give presentations on the work performance and new product ideas of the branch, which received a lot of recognition. It always felt positive when I communicated with then-CEO Eric Schmidt as well as the two founders, Larry Page and Sergey Brin. I was always able to get the kind of understanding I needed from them.

During my four-year term, Google gave its China team more freedom in product development than other international corporations did to their China branches. I greatly appreciated the headquarters' trust, which gave Google China enough space to grow creatively, especially knowing most companies wouldn't do the same.

After taking one more glance at my fond memories of the past four years, I pulled my thoughts back to reality. I was about to have a totally different type of conversation with my immediate supervisor, Alan Eustace.

As senior vice president of Engineering and Research, Alan takes charge of Google's most valuable asset, more than 10,000 engineers. He was the one who called me in June 2005 to say, "Kai-Fu, I've got you an offer I believe you cannot refuse."

Alan is a tall man five years my senior, with a little gray in his short brown hair. He always keeps a friendly smile on his face and a mellow look in his blue eyes.

"Hi, Kai-Fu!" Alan greeted me. "We haven't seen each other for quite a while. How have you been?"

"Pretty good!" I said. "Everything's going well. How about you, Alan?"

We started a little small talk as usual. After a while, I seized a transition point of our conversation to say, "Alan, there's something I need to tell you."

"Oh?" Alan's facial expression suddenly became more serious. "What is it?"

"Alan, this is something I've thought about for quite some time," I said slowly, hoping my calm tone was reflecting the deliberate nature of my decision. "While I do enjoy working for Google, and I do appreciate the headquarters' strong support of Google China, I feel I still have a dream that hasn't been realized. I'm proud of having worked for Google, a company that has changed the world and continues to benefit mankind. But in the next stage of my life, I'd like to focus on realizing my own dream. I'm here to submit my resignation to you in person."

"Really? Are you sure?" Alan seemed shocked. "You know we'd like to renew your contract. We held a meeting in April just to discuss the stock offer in your next contract. Four years ago, we gave you a record breaking amount of stock to make up for your loss of Microsoft stock. Now we are going to give you another large amount, to show our satisfaction with your excellent performance and appreciation of your hard work. We thought you were definitely going to stay with us. Kai-Fu, is there something in the contract or at work that makes you unhappy? What's wrong?"

"Nothing. Really, there's absolutely nothing wrong," I assured Alan. "Google is the most amazing company I've ever worked for, and I've learned a lot in the past four years. This is not about Google at all. It's purely about me, about what I want to do next with my life. Actually I was going to tell you in June. But you know, there was an urgent problem with Google China in June. I had to take care of it. I reminded myself of my responsibility, and told myself that I couldn't leave until after solving the problem. So, I waited until now to resign. Now everything is back to normal at Google China. I can leave without concerns!"

Alan stopped smiling. He slightly frowned, "Kai-Fu, would you please listen to me? I'd like to ask you, could there be anything, any reason or any kind of offer that would keep you? You just started to take charge of our Southeast Asia and Korea teams. We'd very much like to expand your work area. Would anything make you reconsider?"

"Thank you, Alan. But I'm really not here to ask for a higher position or salary. The existing offer is already very generous," I explained. "I choose to leave now because everything is on track with Google China. I can leave with no regrets. There's only one regret in my life now, and I'd like to make up for it. I'm planning to establish a new venture that helps young people in China start up their own businesses, to turn their innovative ideas into useful products."

"So, you are going to start your own company?" Alan gave me a puzzled look.

"Yes, I'll create a platform for China's new start-ups," I said firmly. "Frankly speaking, Alan, I feel I must start right away. I'm already middle-aged. If I don't take action right now, I'm afraid it'll be too late."

Alan stopped trying to change my mind. But in his silence he seemed to be murmuring to himself, "Is Kai-Fu crazy? Why would anybody make a crazy decision like that?"

## Age and Courage

"Are you crazy?" some relatives and close friends of mine directly uttered what Alan might have felt like saying to me.

"You must be kidding! What can be better than working for Google?"

Those were typical responses I received after telling everyone about my resignation. Indeed, I had a position thousands of people would kill for, with the most desirable employer, in the most pleasant working environment, surrounded by the most creative engineers. What more could I want?

In fact I didn't want to gain anything more. I just wanted to do more, actually to give more.

Looking back upon my work history, I realized my experiences at Apple, SGI, Microsoft and Google had cultivated some special energy in me. The energy often propelled me to come up with all kinds of new product ideas. I frequently caught myself imagining the design of a dream product in the middle of my work day, and forced myself to return to my busy schedule at Google China. I longed for a lot of free time, which I could spend planting the seeds of my creative ideas and watching them grow. I also hoped to help other inventors incubate their creations. I could offer them guidance as an experienced mentor. Ideally the platform provided by me would generate as much power as Harry Potter's magic book to change and improve something in the world.

I decided to name the unprecedented new venture Innovation Works, for which I had to leave Google. I had to give up everything I held dear at Google China. While feeling attached, I told myself to let go, and convinced myself that only letting go of the present glory would enable me to go ahead, striving toward an even brighter future.

It was a tough choice, probably the toughest one in my whole life.

Choices in life tend to put more and more at stake as one grows older and takes on family responsibilities. It definitely requires more courage of a middle-aged person than of someone young to take a risk.

In most cases, people avoid taking chances after a certain age. The older they get, the more they have to lose. They are therefore more scared of losing. They would rather keep the status quo, whether being happy or not, just to play it safe.

I, on the contrary, hold an exactly opposite point of view. I believe in taking more courage while aging, because there is less time left.

As I told Alan, I was afraid it might be too late if I didn't immediately start pursuing my next goal.

I felt increasingly pressured by the limited length of life after the passing of Randy Pausch, the Carnegie Mellon University professor who became world famous for his filmed speech, "Last Lecture---Really Achieving Your Childhood Dreams," on YouTube. Randy had been my classmate at Carnegie Mellon. We both completed the Ph. D program in Computer Science in 1988 and attended the same graduation ceremony. Today I still keep a photo of our class waiting in line to receive our diplomas. In the picture, there is only one person standing between me and Randy.

*Randy Pausch (left) and me at the same Ph. D graduation ceremony in 1988*

No one on that graduation day knew Randy only had 20 more years to live. None of us can predict how much longer we are going to be around. How can we not seize the day?

"The key question to keep asking is: Are you spending your time on the right things? Because time is all you have," said Randy in his "Last Lecture."

Before Randy left the world, he also advised us not to fear obstacles. He said, "Brick walls are there for a reason. The brick walls are not there to keep us out. The brick walls are there to show how badly we want something, because the brick walls are there to stop the people who don't want something badly enough. They are there to keep out the other people"

When I watched "Last Lecture" on YouTube, I was amazed to realize how much Randy and I were thinking alike. I should have become closer friends with him at Carnegie Mellon. It was a regret that my busy research kept me from getting to know my Ph. D classmates well.

Although I hadn't been very close to Randy, his death from pancreatic cancer in July 2008 utterly saddened me. He was 47, only one year my senior. What a brilliant but brief life!

I was 47 when making the decision to leave Google for founding Innovation Works in summer 2009. I didn't take it for granted that I outlived Randy. I was going to make every day of my life a step closer to my goals.

## Leading Our Lives

I really liked the line "Lead your life" in Randy's "Last Lecture." It's interesting that he said, "Lead your life" instead of "Live your life." What's the difference between "lead" and "live"?

In my view, to live a life is to simply go with the flow, but to lead a life is to actively shape one's own destiny.

How much control do we have over our destinies? Some may say fate is pre-determined. But based on my experience, I can attest that a lot of it is in our own hands.

Only accidents, diseases and wars are beyond our control, despite certain measures we may take to prevent them. Most of the time, when we are safe and sound, we have a choice of directions. There are many paths in life, and the one you choose right now will take you where you are going to be later. I am where I am because of the choices I have made. I have realized most of my dreams. Now, I feel it's time for me to help others achieve theirs.

I remember Randy on YouTube made me smile when he said, "As you get older, you may find that enabling-the-dreams-of-others thing is even more fun." Perhaps it is because I'm getting a little older. I do find enabling the dreams of many young people even more fun than fulfilling my own wishes.

Innovation Works is a dream factory I have created for them. I have made it very clear in the mission statement of Innovation Works why such a dream factory is greatly needed in China:

*The Chinese entrepreneurial environment is still in its formative stage, with significant barriers for the early-stage entrepreneur: the lack of management experience and coaching, the reluctance of venture capitalists to invest in companies in the formation stage, and the lack of networking and experience to pull a company together. These barriers all contribute to a dearth of high-tech start-ups in China. Innovation Works is matching entrepreneurs, engineers, ideas, and capital with a unique business model that improves success rates and speeds time-to-market.*

*Innovation Works will be the de-facto institution for launching the most promising technology ideas in China. Through the rigorous development and testing of prototypes, and identification of a 'founding executive' to lead the venture, Innovation Works will provide capital, manpower, legal, financial and IT support. Our commitment is to mentoring and supporting the next-generation of Chinese entrepreneurs so that they can focus on building great products without distraction.*

Founded in September 2009, Innovation Works incubates new Chinese high-tech companies and mentors next-generation Chinese entrepreneurs. We specialize in Internet, e-commerce, Mobile Internet, and cloud computing. Every year we plan to prototype around 20 new ideas and spin off about five independent companies.

Figuratively speaking, Innovation Works is a match maker. We bring great ideas, great engineers, great entrepreneurs and great venture capitalists together.

As our unprecedented business model addresses a pressing need of China's high tech market, we have attracted more talents than I have ever imagined. On the first day of Innovation Works, I saw more than 7,000 resumes in our email in-box.

Compared with the 3,000 resumes I received on my first day at Google China, and the 1,000 resumes in the beginning of Microsoft Research Asia, 7,000 is an astonishing number. It has given me a strong boost of confidence in the new venture's future.

We still have a long way to go. But we believe we will eventually get where we want to be. We have patience and passion, the most crucial elements of achievement.

I have transformed from a senior executive of a large corporation into a start-up founder and career coach. I stand by young entrepreneurs to give them guidance and help them hone ideas, recruit people, and secure financing. It is exceptionally understanding of our investors not to expect quick returns, so we can take all the time we need to incubate our projects, to keep nurturing them until they are truly ready to spread their wings.

Before each spin-off, I will give the entrepreneurs my favorite quote from Steve Jobs:
"Your time is limited, so don't waste it living someone else's life. Don't be trapped by dogma - which is living with the results of other people's thinking. Don't let the noise of others' opinions drown out your own inner voice. And most important, have the courage to follow your heart and intuition. They somehow already know what you truly want to become. Everything else is secondary."

Since Jobs lived by all the inspiring words of his own until the last minute of his life, his quotes are now even more powerful than ever.

As a former Apple executive, I lament deeply upon Jobs' death from pancreatic cancer, the same disease that killed my former classmate Randy. When I first heard the sad news, I was also told that a rainbow appeared over Silicon Valley the day Jobs passed away. To me, that sounded like an encore performance by a magician.

Jobs was a magician. He went beyond an engineer's capacity to take the general public into a new world. While there are many distinguished high tech entrepreneurs in our era, he was the only one who made stunning breakthroughs in such diverse fields as computers, operating systems, telecommunications, music and animation.

Without Jobs, today's world wouldn't be the same. Without Jobs, Apple II wouldn't have come out in 1977. Nor would Macintosh in 1984, iMac in 1998, iPod in 2001, iPhone in 2007, and iPad in 2010 have made their grand entrance.

As technologies continue evolving to outdate current products, I think someday iPhone, iPad and Mac may bow out gracefully just as the magician himself did. But no matter what new electronic products future generations will be using, they will always find jobs' speech of the 2005 Stanford University Commencement appealing, and will continue to quote his "Stay hungry; stay foolish."

Now at Innovation Works, besides quoting Jobs, I always bring up my past failures and mistakes to help young Chinese entrepreneurs prevent the same types of problems. I will assure them it's not the end of the world if they make mistakes. Failures are only learning experiences that will build up to their future success.

Since China has a face-oriented culture, many young Chinese people are surprised that I often talk in public about the mistakes I have made. They enjoy listening to the stories of my life not only for their eventfulness but also for their truthfulness. Whenever I mention something of my past, they always urge me to tell more. And I have been a little astounded to hear them ask me as much about my childhood as about my career history.

What intrigues them about my childhood? I guess they want to know how my upbringing has contributed to what I am today. That's why many biographies start with childhood episodes.

In my case, however, I will go back further in time, to an era in which I didn't exist but there were determining factors of my later existence. To me, that was my true beginning.

# CHAPTER 2

# Adventurous Genes

My story began long before my birth. My adventurous spirit emerged in a 12-year-old girl three decades before I was conceived. It made the girl jump on a train, going far away from home to a new world.

The year was 1931. Japan had occupied three northeastern provinces of China, renamed the area Manchuria, established a puppet government to rule it and called it a new country. The puppet government mandated every school to teach Japanese as the official language and put up a portrait of the Japanese emperor. All the students in Manchuria had to bow to the Japanese emperor's portrait every morning before they started classes.

Many college and high school students in Manchuria resented becoming second-class citizens. They also yearned for learning more about Chinese history and reading more Chinese classics. To get the kind of education they desired, they fled to areas of China that were still under Chinese rule. Those who made it would receive financial aid from the Chinese Nationalist government to attend Chinese schools. As they wrote home about it, more and more students followed their steps. That became a trend. But most of those influenced by the trend were over age 15. The 12-year-old girl everyone called Ya-Ching was the youngest among them.

Ya-Ching precociously understood the plight of the Japanese occupation, partially because she was the youngest child of the Wang family, with knowledge beyond her age acquired from older siblings. However, she wasn't following an older sibling when she left home. She departed alone.

# A Union of Love

Ya-Ching arrived in Beijing. She spent the next six years in junior and senior high school there. Then she passed an entrance exam to a junior college that specialized in physical education and was located in Shanghai. That meant she would move further south, going even farther away from her hometown.

Again, she hopped on a train alone. But by this time she had grown into an attractive young lady. Her presence on the train caught the attention of a few gangsters hanging out on the platform. As she sat by a window of the train, snacking on scoops of a half watermelon, they pointed at her and exchanged opinions about her from a distance. She felt a little annoyed but didn't respond until the train began to move. Then she waved at them. When they rushed to her window, she turned the half watermelon in her hand upside down and dropped it on one of their heads. Before they were able to react, the train had taken her away!

Decades later, she told the story to my daughters, making them laugh out loud. At that moment I saw youthful sparks in her eyes, which enabled me to imagine her blossoming years more vividly than the black-and-white photos she had shown me.

One of the black-and-white photos was taken in a professional studio, in which she had her hair all curled up. The picture was such eye candy that the photographer displayed an enlarged copy of it as his advertisement, which allured many male college students walking by. They found out who

*Ya-Ching in the 1930s*

she was and went to the campus of her school just to take a look at her, like fans following a movie star. But they didn't dare to approach her due to the conservative Chinese customs at the time. Despite a large number of secret admirers, she never had a boyfriend until she met the love of her life.

In the winter of 1938, Ya-Ching went to a seminar designed for future teachers and held at a government-run training center in Xian, a city in northwestern China. The seminar included speeches on current events because it was war time. As China was fighting hard against the Japanese invasion, officials of the Chinese Nationalist government often gave speeches to inspire patriotism in young students.

One of those speeches utterly captivated the 19-year-old Ya-Ching. She found the speaker on stage incredibly charismatic. He was average-looking, and a little on the short side, definitely not the tall, handsome type that would attract most girls. But the passionate sparks in his eyes lit up her eyes. His firm, masculine voice with an accent of China's Sichuan Province aroused a tender, feminine reaction from her heart, which she had never felt before. She was unable to turn her focus away from him even after his speech. She looked at him walking away, feeling a strange sense of loss for the stranger...

All she knew of him was his family name Lee, his given name Tien-Min and his head counselor's job title at the training center. She thought she would never see him again after the seminar. However, it turned out that one of her friends in junior college was married to one of his friends. They met again through the couple and fell in love.

He was 10 years older and had a past. He told her about his beloved late wife and two small children. She didn't mind. He also told her about his career path. He first joined the nationalist army when he was a 13-year-old boy slightly shorter than the length of a rifle. Later, he was admitted into Huangpu Military Academy, which was China's equivalent to West Point, in its 6th class. But before he graduated from Huangpu, he received financial assistance from a relative to study abroad. He spent the next five years in Japan and earned a Bachelor's degree in economics there. After returning to China, he first worked as a newspaper editor in Nanjing, the capital of the nationalist government, and became well known for his superb writing. Then he taught at a military school in Chengdu, a city of Sichuan Province, before taking the head counselor's position in the Xian training center where he met her.

Ya-Ching admired him for being a self-made man and related to him for his being a brave soul like herself. The soul mates married in 1939. He brought her back to his hometown in Sichuan Province to live with his mother and children.

The six-year-old girl and a four-year-old boy missed their deceased mother terribly, so it was hard for them to accept a young step mother. In the meantime, the 20-year-old Ya-Ching had to follow all the strict rules set by her traditional mother-in-law. But she didn't complain a word because she loved her husband deeply. Love made her willing to share all of his baggage.

With kindness and patience, she gradually won over her step children while becoming a mother. She gave birth to three girls in the 1940s, the most dramatic decade of China's 20th century. As soon as the Japanese were defeated at the end of World War II in 1945, a civil war broke out between the Nationalist government and the Communists. The war went on until the Communists established a new government in Beijing and the Nationalist government fled to Taiwan in 1949.

The 30-year-old Ya-Ching realized her husband had to leave for Taiwan, or the Communists would kill him for being a legislator of the Nationalist government. But it would slow him down to bring a big family along, so she let him go alone. She stayed behind in order to take care of the family.

Many Chinese families were separated the same way that year. The father working for the Nationalist government went to Taiwan. The mother and children remained in mainland China. In most cases, the separation continued for more than three decades until the two Chinese governments began to allow unofficial communication in the 1980s. That could have been the case for the Lee family, and in that case I would have never been born.

It didn't happen that way, because the adventurous spirit that had once led the 12-year-old Ya-Ching out of Manchuria came out again. It made the 31-year-old Ya-Ching decide to leave for Taiwan, taking all the five children with her.

She made the decision in time. The Communists tightened restrictions on travel later. But in 1950, it was still possible to leave mainland China for Hong Kong while there was absolutely no direct transportation between China and Taiwan. Hong Kong as a British colony then could work as a stop-over.

To go to Hong Kong, the young mother and five children first had to take the train to Guangzhou, the Chinese city closest to the then-British-colony. On their long train ride, the police often stopped them and searched through their belongings. The only boy of the five children had hidden a small nugget of gold in a flash light to prevent it from being confiscated. But one of the policemen almost opened the flash light to check what was inside. He didn't only because the youngest of the five children called him "Uncle" and smiled. He was distracted and forgot about the flashlight.

The gold paid for the expenses of the mother and children in Guangzhou and Hong Kong for months, until they finally reached the father by phone and reunited with him.

## The Naughty Baby of the Family

Before I became the baby of the reunited Lee family, my parents had one more daughter in 1953. It was not easy to raise six children with my legislator father's fixed salary and my PE teacher mother's small income. My parents didn't want more children. No one expected my mother to accidentally get pregnant in 1961.

My mother was already in her early 40s. Pregnancy could be tough on her middle-aged body. Doctors also said babies from older parents were at higher risks of birth defects.

"What if the baby has Down syndrome?" "Isn't it too much of a financial burden to have one more kid?" These were typical comments from relatives and friends on my mother's new pregnancy.

"Are you sure you want to keep this baby?" They asked.

"Yes, I will," my mother replied in a low but firm voice. The adventurous spirit in her manifested itself again.

My mother couldn't explain why she believed I was going to be a very healthy and very bright child. She just simply felt it, and the gut feeling kept her from worrying about all the possible risks of which everyone was warning her.

In December 1961, I was born, as perfectly healthy as my mother had foreseen. The entire family was thrilled. They regarded the new baby as a gift of surprise from Heaven.

*Baby Kai-Fu on Mom's lap with Dad (front center), older brother (right in back row) and older sisters in 1962*

*Excited about my first birthday cake with my father standing behind*

*With my third sister's help to blow candles on my fourth birthday cake*

*Cutting my fourth birthday cake*

I was literally the baby of the family, not only the youngest but also far younger than any of my siblings, even a year younger than my eldest sister's son Wei-Chuan.

My family made Wei-Chuan call me "Uncle," but this slightly older nephew actually behaved more like a brother. He was my closest playmate in early childhood.

Everyone said I was naughtier than Wei-Chuan and perhaps the naughtiest child they had ever seen. I always imitated people I found interesting, including TV characters. I particularly enjoyed putting on my father's Sichuan accent to talk and took big, slow steps like his to walk.

I told my mother I wanted to be a soldier, so she had a tailor make a military uniform for me. But I complained that the uniform didn't have stars on it. Then my second sister asked a general her friend's friend knew for a few stars and gave them to me. I loved wearing the uniform with the stars and holding a toy gun. Every night I made my second sister play a game called "The Soldier Catching the Gangster" with me. I was always the soldier that killed her, the gangster, in the end.

I liked running around so much that it was nearly impossible to make me sit for more than 10 minutes. Every time I needed a haircut, my mother would take my third sister along. Then both of them would perform a puppet show for me in the barber shop. That was the only way to keep me sitting still until the barber finished cutting my hair.

In addition to being extremely active, I was also a rebel. When I was told something, it only made me want to do the opposite. Since my mother often said swallowing a piece of gum would damage the stomach and intestines, I purposefully did it to prove I was as invincible as some cartoon characters I admired. When my mother told me to keep the gum away from my hair, I intentionally attached it to my head. But then I was unable to take it off no matter how hard I tried. I couldn't help but take a pair of scissors to cut off the part of hair stuck with the gum. That made me look funny in school for quite some time until the hair grew back.

Even so, I still wouldn't just believe what I was told. I needed to experiment everything.

When a next-door neighbor bragged about having 100 fish in the pond of his yard, I thought I would have to count the fish before I could believe him. Then one day, the neighbor's whole family was out but left the door unlocked. I seized the opportunity by gathering my fifth sister and nephews to take water out of the pond with buckets so we could count the fish in the almost dry pond. We were very happy about proving the neighbor wrong. But we didn't realize we were killing the fish until the neighbor found out and took the case to my mother.

Uh-oh! I got scared when the neighbor came. I thought my mother would punish me. But she didn't do anything to me after apologizing to him. I even caught her trying to suppress a laugh. She seemed to find the incident funny. That was the first time I saw there was also a naughty child in my mother.

My fun-loving mother greatly enjoyed a Chinese board game named Ma-Jiang, which she was only able to play with her friends after I went to bed. But I always had too much energy to fall asleep early. After she put me in bed, I would ask myself, "Why do children have to go to bed so early? How come adults can play at night?"

One night I really didn't want to go to bed, so I came up with an idea to turn back the time by one hour on all the clocks at home. I even did the same to my sisters' watches, which they usually took off after coming home. The time change enabled me to stay up for one more hour. I was extremely excited about it. However, the time change also made all the family members get up an hour late the next day. My sisters were screaming. But my mother didn't scold me. I even overheard her say to my father, "Our baby boy is quite smart!"

# The First Breakthrough of My Life

I made the first major decision of my life when I was five. I told my mother that I was tired of all the songs and games in kindergarten. I asked her to let me go to first grade.

My mother was surprised, because she had never heard of other children doing that. She said, "In a year you'll be in first grade. Why wouldn't you wait a year?"

I continued to persuade her, "Mom, how about letting me take an entrance exam to a private school? If I pass, you'll let me go to first grade. If I don't, I'll go back to kindergarten."

My mother thought about it for a while, and then she agreed.

In Taiwan, public schools are open to all from first to 9th grade, but private schools hold entrance exams. The situation is not exactly the same as but similar to the common practice of schools in the United States.

I was the youngest among the children taking the entrance exam to Ji-Ren Elementary School that year. When the exam results were posted at the school entrance, my mother took me there. She immediately saw my name as the first on the acceptance list. In Taiwan, the acceptance list always starts with the person obtaining the highest total score.

My mother screamed, "Ah! You passed! And you are number one!"

I will never forget how excited she was. That was the first time I saw how a child's small success could make a mother proud big time.

In the meantime, I learned to be bold about breaking through limitations. Now in retrospect, I deeply thank my mother for allowing me the first breakthrough of my life, especially knowing Chinese parents conventionally tend to be on the cautious side. My parents made an unconventional choice of letting me take charge of my future at such an early age. I was incredibly fortunate!

While I started school a year earlier than most children, I soon found out even the first-grade curriculum was too easy for me. The first-grade math only covered addition and subtraction, but I already knew the multiplication table by heart under my mother's tutoring. I was also able to recite many classical Chinese poems my classmates didn't know.

Being ahead of the class made me often get bored in class, so I talked to my classmates or passed messages to them. Sometimes I made faces for fun.

One day, my restless behavior offended the teacher. After giving me a few warnings, she sealed my mouth with scotch tape.

My mother happened to be early that day for picking me up. I was terrified when she saw my sealed mouth. But she didn't scold me.

My mother was a lot more liberal than most Chinese parents of her generation. While they taught their children to be obedient, she never stopped me from challenging authority.

Bolder than my classmates, I would correct the teacher's pronunciation in English class based on the standard American English I had heard from my fifth sister's tutor. That made my classmates laugh in class and look at me differently after class.

I did like to stand out. Once I bragged to my classmates that I had learned special kung-fu and could digest paper. To prove it, I tore off a piece of paper from my notebook and ate it in front of them.

"Wow!" Their eyes all opened wide.

"What else can you eat?" One of them asked.

"I can eat wood," I said. "I'll show you how I can bite my desk."

I did bite into my desk every day during lunch time until I made a large dent on it.

To astonish my classmates, I even claimed I could swallow lead from a pencil, and demonstrated it. That brought me to an emergency room. The doctor gave me a serious warning and prescribed medicine.

I stopped telling people I could eat inedible things. But I still wanted to be different from everybody else. I dreamed of being a hero who would save lives and punish the evil.

I looked around for the evil to conquer, and something did appear on my radar. I noticed that a teacher always fined students who talked in class and claimed the fines would go into the budget for class activities, but the class budget didn't seem to have increased as much as all the fines he had collected. I double checked with the student officer in charge of the class budget, and found out there was indeed a difference. Apparently the teacher had pocketed the money.

For the sake of justice, I wrote a letter to the principal with my left hand despite being right-handed in order for no one to recognize my handwriting. The principal questioned the teacher the next day. After that, the teacher shouted at the entire class, "Don't think I don't know who did it. Your behavior to turn against your teacher is no different from the Communists in the Cultural Revolution!"

I tried to hide my fear and appear calm. I told myself my work was meaningful, knowing no more money for the class budget would go into the teacher's personal pocket. I felt I was becoming a hero like those in the kung-fu fantasy novels I had read.

I loved kung-fu fantasy novels so much that I decided to write one in fifth grade. I collaborated with Wei-Chuan, the nephew one year my senior. It turned out to be a graphic novel, titled *The Mystery of the Animal Fighters,* containing tens of thousands of words and numerous drawings. We marked page numbers to make it look like a real book. I even put a note about the copyright on the back cover, which read, "Published on Aug. 3, 1972/ Anyone who copies this book is punishable by death!"

*Buddies with nephews around my age---from left to right: Xiang-Sheng, Wei-Chuan, me and Yu-Sheng*

The main characters of the book were based on my family members, with everyone's imperfections exaggerated.

My family members all laughed out loud about the book's funny-looking characters and fun-poking descriptions. Even neighbors asked to borrow it after hearing about it. I felt very proud, and my mother regarded it as the biggest literary achievement of my childhood. She still keeps the book today.

## Spoiled, but not Rotten

Everyone said I was my mother's favorite child. My sisters didn't really mind, because they were a lot older and didn't need as much care as I did. However, my fifth sister sometimes would express a little jealousy half-jokingly.

She said, "I used to be the baby of the family. But Kai-Fu took that away from me and turned me into a Cinderella washing his diapers!"

My fifth sister was already eight years old and capable of simple household chores when I was born. My mother didn't have to do much for her, and certainly not for the even older children, so I was naturally the focus of my mother's attention.

Since my mother had me late, her body didn't produce as much milk for me as in the nursing times of her younger years. To make sure I would receive enough nutrition, my mother followed a Chinese recipe that was designed to stimulate lactation. She made a stew of pig feet and peanuts, which she forced herself to eat every day. The recipe indeed worked to provide me lots of milk. But it put extra pounds on my mother at the same time, and probably changed her metabolism, too. Her once-slender shape was ruined because of me. But it didn't matter to her. She cared about me far more than about herself.

My mother treasured me so much that she did more than normal to protect me. She didn't let me take the school bus. She hired a driver to bring me to school every morning when she had to teach PE at a Girls' high school. In the afternoon she was done with her teaching job, so she always picked me up from school.

"I can spot you right away from a distance because of your little white legs!" She often said fondly when approaching me. I had lighter skin than my classmates, perhaps due to less sun exposure.

Whenever my school held a field trip, my mother always wrote a note to the teacher to get me a sick leave, because she was afraid I might fall and get injured in an unfamiliar environment. I felt a sense of loss every time my classmates took a day trip and I stayed home. I also envied my classmates for their getting to eat all kinds of snacks bought from street vendors. My mother didn't let me touch anything sold by street vendors because back then sanitary conditions in those vending stalls were questionable. Even when we went out to eat at a squeaky-clean-looking restaurant, my mother would have me use utensils brought from home. She was doing everything to prevent me from getting sick.

My fourth sister thought my mother worried too much, and she knew I craved street vendors' fried pan cakes, which she considered harmless, so occasionally she would bring some for me behind my mother's back. Whenever my fourth sister showed me wrapped-up fried pan cakes hidden in her bag, my mouth immediately watered. Today I can still almost smell the aroma of Taiwanese fried pan cakes when thinking of them.

Although I rarely had a taste of street vendors' food, I ate a lot. My mother was (and still is) a fabulous cook. My favorite was her mini dumplings with spicy Sichuan sauce. They were smaller than regular Chinese dumplings because my mother placed a tiny tea cup (like those used for the Japanese tea ceremony) upside down on each store-bought flour wrap and divided it into little wraps with the same diameter as the mini cup. Then she put top-quality ground pork, mixed with water to ensure tenderness, on all the little wraps and folded them up. Once they were boiled, they would almost melt in the mouth. The sauce for dipping the mini dumplings also came from my mother's secret recipe, which contained home-made chili oil, chili pepper, peppercorn, and minced garlic. My mother said the timing of adding the garlic would have to be right to ensure the best taste.

Beef noodle soup was another great dish my mother often made for me. Years later, I introduced its recipe to the cafeteria of Google China, and it became one of the most popular dishes there. The beef noodle soup had the spiciness of my father's home province Sichuan but was more flavorful than most Sichuan soups. The beef was particularly tender because it came from a cow's front legs. Beef from a cow's rear legs just wouldn't taste the same!

My mother's crispy pork was also more tender than usual. She would bread a well selected pork chop, fry it, slice it and steam the slices with special Chinese vegetables. When it was done, it looked and smelled just as appetizing as it tasted.

Every day after school, my mother would ask what I felt like for dinner, and then that would be what the whole family would eat.

Her typical question was, "What should we eat tonight? Baby boy?"

My most frequent answer, "Mini dumplings with spicy Sichuan sauce!"

"OK," she smiled and put on her apron, looking very happy to start making this time-consuming dish for me. "How many are you going to eat?"

"Forty!" I replied cheerfully, looking forward to eating more dumplings than everyone else in my family.

Overeating every night made me the heaviest kid in class. Especially in fifth grade, my weight skyrocketed. Today no one can believe I was once that much overweight. But I indeed was.

My mother became concerned about my weight. She would say at dinner, "Well, you've already had a lot tonight, and you are kind of too fat. Stop eating now, OK?"

"Oh! Just let me take one more bite," I couldn't give up the delicious food in front of me and would beg her. "Last bite before leaving the table, please!"

After dinner, I always put my homework aside to watch TV. By the time TV programs ended, I was already too sleepy to do homework. Then my mother let me go to bed, and she would wake me up at 5 a.m. for my homework.

My mother also gave a wake-up call at 3 a.m. to my fifth sister, who wanted to study for her college entrance exam in the quietest hours. Both of us counted on our mother more than an alarm clock, and she really never missed once!

I didn't know how it felt to get up at 3 a.m. and then again at 5 a.m. until I had a baby daughter and began to change her diapers in the middle of the night. Then I truly realized how much my mother had sacrificed for us..

Today's child psychologists may say my mother was overprotective, and I admit my mother overprotected me, but I wasn't spoiled rotten. My mother had a strict side when it came to my education.

I kept receiving perfect scores in the first few weeks of first grade. One day a friend of my mother's visited us. She casually asked me, "How's school?"

"I get 100 on every test," I bragged. "I've never seen a 99 and don't even know what it looks like!"

But soon after that, I received a 90. My mother saw the score and spanked me (when I was little, it was acceptable and even expected in Chinese culture that parents exercised physical punishment to discipline children ).

"Why hit me for a 90? That's not a bad score!" I protested.

"This is not for the 90 but for your bragging," my mother explained. "You shouldn't have bragged about getting 100 every time even if you could keep it up, not to mention it's not necessarily the case! Keep in mind modesty is a virtue we Chinese people emphasize. Don't ever brag again!"

My mother held high standards for my learning, and enforced her requirements with carrot and stick. She made me memorize important passages from my textbooks and recite them to her. If I missed a word, she would throw the book across the room and tell me to pick it up. Sometimes she spanked me with a ruler when I made too many mistakes. Once she broke the ruler when hitting too hard. So, the stick part was practically literal. As for the carrot part, she bought me a present for every number one I obtained.

Once I asked for a Chinese version of *Sherlock Holmes* as a reward for my outstanding grades. My mother bought not only the series of books but also a watch for me. She said, "Books don't count as a reward. We can buy books any time you want."

With her encouragement, I read hundreds of extracurricular books a year. I read Chinese translation of Western masterpieces, such as *A Tale of Two Cities* and *The Count of Mount Cristo,* as well as Chinese classics like *The Three Kingdoms* and *The Monkey King.* My favorites were biographies, especially of Helen Keller and Thomas Edison. I admired Keller's courage to overcome her physical disabilities and Edison's inventions that changed the world. I began to desire becoming a scientist.

Like Edison in his childhood, I sometimes became absent-minded when thinking about a puzzling question. That kept my exam scores from being always near perfect. Once I got a 78, the lowest ever. I was terrified, picturing how my mother was going to hit me. Then a thought popped up when I saw how similar the 78 looked to 98. I used a red pen to change the number and showed it to my mother, with my heart pumping to my throat. She didn't find out.

The next time I received a lower-than-expected score, I tried to do the same, but my hand trembled and messed up the number. I was horrified. There was no way I could bring this back to my mother, so I threw it into the garbage can.

My mother didn't ask me about the exam that day. She seemed to forget about it. But the secret kept bothering me. A few days later, I finally mustered all my courage to confess to her.

I thought my mother was going to hit me hard. But she simply said, "I'll let it go as long as you know you were wrong. Just remember to be always honest from now on."

That was one of the most important lessons my mother taught me. Thanks to her, I constantly hold honesty as one of my core values.

# My Father's Influence

I was not as close to my father as to my mother. Like most traditional Chinese fathers, my father always kept a serious image in front of the kids. He didn't play with me. When he was home, he spent most of the time in his study, with the door of the study ajar. I often saw him sitting in front of his desk, concentrating on his writing. At those moments he somehow seemed a little mysterious to me.

My father had a heavy accent of his home province Sichuan. We usually spoke the Sichuan dialect with him while speaking Mandarin, the official Chinese language, with our mother. Today I still remember a lot of the Sichuan dialect.

My father was not as expressive as my mother. To me and my siblings, our mother was as warm as sunshine and our father as calm as moonlight. The sunshine was so strong that it made the moonlight seem invisible in its presence.

I didn't feel my father's love throughout my childhood. Now, with more understanding of Chinese traditions, I've realized he had a more subtle way of expression like many traditional Chinese men. For instance, he would ask me to walk with him when he took a walk, so we would spend some time alone. It's a regret that I didn't understand then.

The closest I ever felt to my father was the moment I solved a math problem he had given me, with his Parker gold pen promised as the reward for the correct answer. The question was how to make four equally sized triangles with six matches. I was only five then, so my father was most surprised to see me come up with the right solution right away. He immediately gave me his Parker gold pen, which was a luxury item in Taiwan then, definitely something too expensive to give a kid. But my father insisted on keeping his promise.

My father influenced me through more actions than talks. He only scolded me once.

That was when I was in fourth grade. I saw a street vendor selling cartoon pictures near school and many students buying them, so I wanted to start a similar business. I shared the idea with Wei-Chuan. He liked it, too. But we needed money to buy the cartoon pictures we were going to sell. For the funding, I took a few thousand Japanese yens from a drawer in my father's study. I thought my father wouldn't notice the loss because he wouldn't think of those yens until planning to go to Japan again, which I knew definitely wouldn't happen soon.

Wei-Chuan and I took the yens to the bank, trying to convert them into the Taiwanese currency. But the bank teller refused to provide service for us because we were minors.

With the business idea failed, I wanted to put the yens back in my father's drawer. But the drawer was locked, so I decided to bury the yens in the back yard and pretend nothing had happened.

However, my parents found out about what Wei-Chuan and I had done. My father told my mother that he would deal with me. I was petrified. I was always more scared of my father than of my mother because he looked colder and stricter.

To my surprise, my father just took me into his study. He calmly said to me, "I hope you won't disappoint yourself again!" Then he walked away.

Somehow those simple words were more powerful than any harsh blame or elaborate lecturing. That understated one line brought up a sudden sense of loss in me and made me reflect upon myself. Then it stayed within me, coming out every time I was going to make a decision. I haven't disappointed myself or my family ever since.

As time went by, my father's influence gradually became part of my soul. I didn't get to know him as well as I wish I could have because I left Taiwan for studying in the U.S. at age 11. But through reading his books and listening to people talk about him, I understood him more and more.

My father was first elected as a legislator in his home province Sichuan in 1948. After moving to Taiwan in 1949, he still worked as a legislator. He was a government official for most of his career, but he disliked politics. There were a lot of under-the-table activities, and some politicians got rich. My father was always completely clean. When he decided to buy a house in his 70s, he actually did not have enough money, and his children, including me, chipped in to buy a house for him.

Starting in the 1960s, my father enjoyed his side job as a part-time university professor and researcher more than his day job as a legislator. He spent most of his free time writing. His goal was to leave an accurate historic record of what had happened in the first half of the twentieth century. His books were banned in Taiwan and China, because they told the truth, instead of blindly following the official versions of history books. His books were only published in Hong Kong and Japan. Today, China and Taiwan are much more open to his writing. But unfortunately, most of his books are out of print already.

My fifth sister was my father's favorite child. After he passed away, she wrote an emotional piece in memory of him. It describes how our father missed his hometown and his mother. According to the article, my father couldn't help weeping every time he heard old songs that had been popular in mainland China in his younger years. The only TV program he watched was a show that introduced various scenic spots in mainland China.

My father wrote the same couplet for every Chinese New Year to express his wish for returning to mainland China. Like couplets in English poetry, a Chinese couplet consists of two lines in parallel structure. There is a Chinese New Year's custom to write the two lines of a couplet on two separate strips of red paper and post them on each side of the door (given the way Chinese writing goes vertically from top to bottom) . The New Year's couplet usually contains auspicious words as wishes for peace and prosperity in the coming year. But my father's couplet for Chinese New Year's was different:

*Day by day I train with my riding horse*
*Year after year I long for a home-returning trip*

It was my father's biggest regret in life that his mother remained in mainland China, so he wasn't there when she passed away.

My father went back to visit his hometown at age 81. It was an incredibly emotional trip to him. The night he returned to Taiwan, he showed my mother and my sisters a large stamp carved by a famous artist in Sichuan. The first line of a classical Chinese poem was engraved on the stamp. It reads, "The young man who left home returns as an elderly." Each time he told the story, he would sob uncontrollably...

One of my father's former students wrote an essay in honor of my father. He said he and his classmates all looked up to my father for being a hard-working scholar. My father spent at least one day a week in the library of the Taiwanese university where he taught part-time. He went to the US every summer to do research at Harvard University and Princeton University.

My father hardly knew English before age 50. But for the research he wanted to do in American libraries, he worked very hard to learn English. He memorized new words every day, watched American movies, and practiced conversations with native speakers. He managed to become communicative in English within two years. Although my sisters and I often made fun of his heavy Sichuan accent when he spoke English, we admired his hard work.

My father's values have become building blocks of my value system. Every time I went back home after he passed away, I would spend some time in his study to appreciate his legacy. He was a quiet father, but only after he passed away did I realize that he taught us not by his words, but by his deeds.

There are two frames of Chinese calligraphy that form a couplet on the wall of his study. Many Chinese intellectuals like having frames of couplets on their walls all year around. Unlike the Chinese New Year's couplets, the framed couplets usually function as mottoes.

My father's motto can be roughly translated as:

Only with magnanimity can you absorb the greatest virtue
*Simply without desire will you build the highest character*

I think these two lines most vividly describe my late father.

*Sitting in front of my father's motto*

## CHAPTER 3

# Flying to America

In the 1960s, there was a trend for Taiwanese college graduates to pursue graduate studies in the United States. My elder brother Kai-Lin was one of those students. He was accepted to Tulane University with a scholarship, and went to Louisiana by ship, as he could not afford an airfare, but a friend offered to take him on a freight ship without charge. When he got off the boat in New Orleans, he only had 10 dollars in his pocket.

Kai-Lin lived frugally on his scholarship. He was financially unable to come home during summer or winter breaks, so he sent photos to us instead. After earning his Ph. D., he became a research scientist for the Oak Ridge National Labs, which helped him get permanent residence and eventually citizenship of the United States. He met his match, Ten-Ching, also a research scientist. They married in Tennessee.

When he brought his wife back to Taiwan in November 1971, he had been away from home for nine years.

*My sister-in-law and brother (front row from left) sitting beside my parents with my four sisters and me standing behind in 1971*

Everyone in my family was excited about his return. We filled him in on all our major events in the past nine years. My second sister had married. She and her husband were fresh college graduates who didn't make enough money to get their own place, so they were living with us. My third sister had graduated from a nursing school and left for further studies in America. There wasn't much to report about my fourth sister, fifth sister or me. But in Kai-Lin's eyes, I had changed most drastically. The baby he had said good-bye to nine years ago was now a big boy!

Kai-Lin asked me about school, and then he told my parents that Taiwanese education was too exam-oriented, too rigid for me. He said American education was more inspirational, more enlightening, and therefore more suitable for a smart, naughty kid like me. He suggested taking me to America.

The suggestion aroused mixed feelings in my mother. By then she had hardly let me out of her sight. It was almost impossible for her to imagine letting me go thousands of miles away. However, she wanted the best for me. She believed an American education would give me a bi-cultural advantage, which would benefit my future career and could even enable me to change the world!

I didn't really know what it meant to study in America. But I was curious about it. The snow in some photos of my brother and sister-in-law intrigued me, as I had lived my whole life in a semi-tropical city where it never snowed. I also really liked a toy tiger my sister-in-law had sent me from America. I held it in my arms almost every day. That made me imagine America to be a fun place with lots of toys.

"You should go to America," my mother said to me. "Many outstanding people were American-educated."

With her consent, Kai-Lin filed paperwork to petition for my immigration.

After I received my green card, my mother and I got on a plane in early December in 1972. My mother was only going to spend six months there to help me adapt to the new environment.

Other family members saw us off at the Taipei airport. We took a family picture there before the flight.

While flying across the Pacific, I didn't really feel I was leaving home. My mother was with me. I didn't think the new life in America would be much different from my days in Taiwan. I slept well with sweet dreams on the flight.

We landed in San Francisco, where we visited my father's friend Mr. Cheng, a professor of San Francisco State University, and his wife, who was working for Stanford Hospital. We stayed at their house for a night before flying to Tennessee.

Oak Ridge is a small town in Tennessee. It used to be an oak forest. In 1914, a man named John Hendrix claimed he saw the future after taking a nap in the forest. He said, "A voice told me that this forest will be replaced with houses and factories, which will help America win a war. After the war, there will be a city here."

During World War II, the U.S. government established a research center in Oak Ridge as part of the Manhattan Project. No one knew what the research center was doing until two atomic bombs exploded in Japan in 1945. Then Oak Ridge became well known for being where atomic bombs were developed.

Hendrix's prediction indeed came true!

After World War II, the U.S. government named the research center the Oak Ridge National Labs. In 1949, the area around it was named Oak Ridge, which officially became a city with a council and mayor in 1959.

When I arrived in Oak Ridge, it was a scenic quiet town. Coming from Taipei, a densely populated city, I felt I was entering an entirely different world.

## My New Life in Oak Ridge

Kai-Lin owned a two-story house in Oak Ridge. I moved into a room on its second floor, with a view of the large yard.

Within the first month of my arrival, I saw snow for the first time of my life. I excitedly ran around the yard, trying to catch snow flakes. As snow accumulated on the ground, I made snow balls to throw at Kai-Lin. We also made a snow man together.

When spring came, colorful climbing roses covered the fence of the back yard. My sister-in-law would cut a few and put them in a vase in the living room.

In summer, I helped Kai-Lin mow the lawn and reap vegetables grown in the yard. The vegetables were the kinds frequently used for Chinese cuisine but not in American food, so they were not available in the local markets. Growing them in the yard enabled us to enjoy some of the same Chinese dishes we used to eat in Taiwan.

My sister-in-law usually cooked Chinese food for dinner, but for breakfast we would have American cereal for its being quicker. My brother and sister-in-law worked very long hours as scientists, so I wanted to help them save a little time. I would get bowls of cereals and glasses of juice ready on the table for them. They were pleasantly surprised to see me serve them breakfast because they knew I had been waited on like a little emperor in Taiwan. I became more independent after coming to America, especially after my mother went back to Taiwan.

Life was very boring for my mother in Oak Ridge. I went to school during the day. My brother and sister-in-law came home even later than I did. My mother would cook for all of us, but aside from that, she spent most of the day watching TV while she knew too little English to really understand the programs. She didn't complain, but her silence told us she was not very happy most of the time. Only weekends cheered her up.

We had more time to be with her on weekends, and sometimes we would have guests over. My brother and sister-in-law had some Chinese friends in town. When they visited on weekends, my mother played the Chinese board game Ma-Jiang with them. That brought her some good times before she returned to Taiwan.

*My mother (right) and me in Tennessee in 1973*

After my mother left, my second sister's son Ray came. My second sister let our brother and sister-in-law adopt her second child because they didn't have children. Ray was only six years old when he moved into my bedroom. He was a shy skinny child wearing thick glasses.

I helped Ray with his homework and taught him many things I knew. But my naughty nature came out to take advantage of him sometimes, too. When we played Blackjack, I tended to get an idea of the cards when distributing them, so I always won. Then I would tell him, "You lost 100 dollars. Remember to pay me later."

Ray didn't know I was tricking him. He wanted to win back what I said he owed, but he just kept losing. His debt eventually amounted to a hundred million!

Right before I left my brother's house for college, I told Ray, "I'd like to give you a present before going away."

"Really?" Ray was surprised. "What is it?"

I cleared my throat and put on a serious tone to say slowly, "I am going to write off your one-hundred-million-dollar debt!"

Both of us burst laughing.

Years later, Ray went to medical school in Washington and then became a successful doctor in Texas. When we get together, we still bring up those sweet six years we spent together.

Winning a State Essay Contest
The first two of my six years in Oak Ridge were a crucial stage of my life. It was during those two years that I became fluent in English, adapted to American culture and built a solid foundation for my American education. I spent those two years in St. Mary Junior High School, a Catholic school run by nuns.

I prayed three times a day with my teachers and classmates. Today I still remember every word of the prayer:

Our Father, which art in heaven
*Hallowed be thy name*
*Thy kingdom come*
*Thy will be done*
*On earth as it is in heaven*

*Give us this day our daily bread*
*And forgive us our trespasses*
*And Lead us not into temptation*
*But deliver us from evil*
*For thine is the kingdom, the power and the glory, for ever and ever,*
*Amen!*

The power of religion touched my heart. I felt a sense of peace and harmony every time I went to church. While attending religious rituals, I gradually familiarized myself with Western culture.

When I first entered St. Mary Junior High School, I knew little English. The basic English vocabulary I had learned from my previous school in Taiwan was definitely not enough for me to cope with the 7th-grade curriculum of America. I had to use a dictionary with Chinese translation to look up new words all the time when reading textbooks.

Listening was more difficult because that wouldn't allow the use of my dictionary. At first I didn't understand any of the teachers' lectures. I got bored and fell asleep in class. Sometimes I brought a Chinese novel to read in class. St. Mary's teachers were so kind that they didn't say anything when they spotted my novel. However, I felt terrible. I hated falling behind after always being ahead of my classmates in Taiwan. I knew the language barrier was the only thing that kept me from being a top student again, so I decided to drastically improve my English in the shortest time possible.

My first strategy was to memorize a thick book of vocabulary. But I soon realized it was not an efficient method. It was very easy to forget a word I had memorized if I didn't use it often.

After seeing the importance of context in vocabulary building, I began to participate in my classmates' conversations. They were all very patient to explain the parts I didn't understand initially. Feeling encouraged, I also became more pro-active in class. When I missed something in the lecture, I would raise my hand and ask the teacher, "Sorry, I didn't follow. Would you please say again what you mean?"

My teachers all had a lot of patience for me and gave me special considerations. They let me take tests at home so I could look up new words, under the conditions that I wouldn't look for the answers in the textbooks. I appreciated their trust and never betrayed it. This enabled me to do well on tests despite my limited vocabulary and boosted my interest in learning. I couldn't thank them enough for the thoughtful arrangement.

I was (and still am) especially grateful to the principal, Sister Mary David, for her going out of her way to help me. She sacrificed her lunch break to tutor me English with a first-grade textbook. Today I still remember the first lesson of the book:

*I have a dog named Spot*
*See Spot walk*
*See Spot run*

We started with such simple sentences and moved on from there for a year. Within that year, I made tremendous progress in English. One day I suddenly noticed that I understood every English word entering my ears. I was ecstatic!

Later I heard about other cases of such success in young immigrants. It seems that those who arrived in America by age 12 tend to pick up English faster and sound more like native speakers than those who came as teenagers or adults.

It was not too surprising that I learned to speak fluent English within a year. What truly amazed everyone was that I began to write better than many of my classmates, who were all native speakers.

Two years after my arrival, I entered a state essay contest that let each student pick a topic related to the biggest challenge America would face in its third century. While most of the students chose to write about the energy crisis or environmental pollution, I decided to discuss something more introspective

*Sister Mary David*

I submitted an essay titled, "Apathy---America's Biggest Enemy in Its Third Century," in which I discussed the trend of hippies in the 1970s, pointed out their apathy to society, and warned against their anti-progress mindset. I concluded the essay with the following sentence:

The biggest challenge American society faces today is nothing more than how to change people's apathy towards one another.

When the outcome of the contest was announced, my entire school was astonished to see my name among the 10 finalists in the state. The state of Tennessee recognized a non-native speaker who had only been in America for two years as one of the 10 best student writers statewide!

The 10 of us were supposed to orally defend our essays, and our oral performance would determine the champion. During my oral section, a teacher wearing thick glasses asked me, "If you consider Americans apathetic, what do you think about Ralph Nader's point of view?"

I couldn't answer the question because I didn't know who Ralph Nader was. That was surely the major reason why I didn't win the championship. I learned from the experience that mastering a language was more than knowing how to express oneself in the language. I realized the importance of cultural knowledge. Then I began to read more and more English books, newspapers and magazines in order to increase my understanding of American culture.

In the meantime, I didn't forget my first language. I wrote a letter in Chinese to my mother every week after her return to Taiwan, and she would correct the word usage in my letters. That helped me improve my Chinese writing. My Chinese vocabulary was expanding at the same time, thanks to all the Chinese kung-fu fantasy and romance novels in my brother's house. I read every single one of those novels at least five times during my six years in that house. That kept me updated on Chinese language and culture.

I was growing up bilingually and bi-culturally.

Differences between the East and the West
In the 1970s, there was not as much global communication as there is today. Small town residents especially knew little about foreign countries. Many people I met in Oak Ridge confused my birthplace Taiwan with Thailand. I often needed to explain the historic relationship between Taiwan and China to them.

My classmates knew I was Chinese. But all that meant to them was just I was someone different. Some of them wondered about the difference and looked for answers in stereotypes of which they had heard.

One day in a PE class, a boy pointed at me and said, "You are Chinese, so you are backward and stupid!"

I felt deeply insulted, but didn't know how to react. At this moment another boy jumped out and talked back to the bully, "How can you call Kai-Fu that? How can you talk garbage like that?"

The two boys started fist fighting. I wanted to stop them and suddenly thought of Bruce Lee, whose movies were popular then, so I shouted at the bully, "Stop! I know Chinese kung-fu. If you don't stop, I'll pull my kung-fu on you!"

I had to correct a stereotype with another stereotype. That made me feel sad. I decided to work on presenting a positive image of the Chinese to mainstream Americans. To that end, I needed to start from myself to win their applause!

I began to display my strengths in school. The easiest way for me to do that was to answer questions faster than anyone else in math classes.

One day, as soon as the math teacher put 1/7= ? on the board, I immediately spoke up, "0.142857!" Everyone was shocked. They didn't know I had actually memorized the answer to the question in Taiwan's elementary school.

Memorization is an essential part of Taiwanese/Chinese education. This actually helps children learn very fast because young brains have good memories. However, too much memorization may make children only know to follow what's in the books. Such children don't get to develop their creative potential or independent thinking.

I was lucky to get the best of the East and the West in terms of education. I first benefited from memorization through my childhood and then developed creativity as well as critical thinking skills as an adolescent.

I was like a fish in water in American education, which let me swim freely and far. I graduated from St. Mary Junior High School with excellent grades and continued to excel at Oak Ridge High School.

My high school math teacher, Ms. Benita Albert, noticed my extraordinary performance in class. She decided to help me advance further by giving me free private lessons beyond grade level.

Ms. Albert also taught part-time at the University of Tennessee. One day she asked me, "Would you like to sit in on my college classes?

Those classes probably suit you better than high school math. I'm sure you'll find them helpful."

Wow! That would be really beyond wonderful, I thought. The only problem was transportation. The college was a little far, and I didn't have a car. Ms. Albert saw through what was on my mind. She said, "I know you don't have a car. That's not a problem. I can pick you up on my way to teach there."

I couldn't believe how nice she was to me! She brought me to her college classes for a year, and I kept making breakthroughs in math that year. A year later, I won the championship of the state math contest.

While going beyond grade level in math, I also challenged myself in English. I took an English literature class that contained Shakespearean language. It was difficult at first. But I soon fell in love with English classics. In addition to Shakespeare, I also read famous English novels such as *Jane Eyre,* and masterpieces of American literature such as *The Scarlet Letter* and *Walden.*

Literature gave wings to my imagination. Although it may not seem relevant to my later career choices of science and entrepreneurship, I believe those literary readings I did in my formative years stimulated my creativity, deepened my understanding of Western culture, and helped make me what I am today.

Best Friends
I was selected along with two other math geniuses of Oak Ridge High School to attend the advanced math program at the University of Chicago in the summer of my sophomore year. The two other boys were Phillip Yoo, an athletic Korean American, and Ram, a handsome mixed child of Indian and Japanese origins.

*My buddies and me (center) in the summer math camp at the University of Chicago in 1977*

I was not close to Phillip or Ram until we went to the math camp together. Before that summer, I always felt I was an outsider in school. I was probably the only one not born in America, so I couldn't relate to my classmates in certain ways and didn't make friends with any of them.

The math camp brought me closer to Phillip and Ram while introducing me to two Caucasian boys from other schools. We formed a group of five, going everywhere together. At night, we stayed up to chat in our dorm room. We talked about our childhoods, our parents, girls we found attractive, and everything else that interested teenage boys. Sometimes we forgot time until the resident fellow knocked on our door and said, "Time to go to bed, kids! It's almost dawn!"

We had so much energy that the lack of sleep didn't bother us. We jumped out of bed early to get breakfast in the cafeteria, where the gray-haired lady cook often shouted at us, "How many eggs would you like? Make up your mind now! Don't change your mind after the eggs are done!"

No one liked dealing with her, so I decided to be the bravest. I told all my classmates that I would get their eggs for them. I gathered all their vouchers and went to the cook, asking for 83 eggs.

"What? 83? Are you crazy?" She screamed.

"Yes, ma'am," I replied calmly.

"You are intentionally giving me a hard time," she said. "I'll report to my supervisor."

When the supervisor came, I explained I was buying eggs for all my classmates. Then he told the cook to accept my order.

I felt triumphant. Although I was already about an adult's size, I still kept my childhood dream of being a hero.

My classmates followed an idea I came up with to sell their remaining cafeteria vouchers after purchase to University of Chicago students at discounted prices. We made a little money, with which we went out to downtown Chicago, ate Chicago's famous pan pizza, and played video games.

We also played bridge. We often hid ourselves in the closed library to play the game until the security guard found out and forced us to leave. During the summer at the University of Chicago, I truly mingled with mainstream Americans and adapted to American culture.

## My First Two Start-up Ventures

After the math camp, I went back to school as a changed person. I became involved in extracurricular activities. I joined the math club and the bridge club. I was even elected vice president of the student committee.

Being active in school inspired me to create a school newsletter. I collaborated with my friends, and wrote anecdotes of funny mistakes our teachers had made as well as satires on certain school regulations we considered unreasonable. The official school paper was called "The Oak Leaf," so we called our newspaper "The Loose Leaf."

I bought an IBM Selectric typewriter and spent two weeks typing up the original version of the newsletter. Then we found a printing factory to work out a deal of publishing a cartoon book by our friend's father, with printing 1,000 copies of our newsletter as a bonus. Since we got 1,000 copies of the newsletter at nearly no cost, we decided to distribute them for free.

As soon as copies of the newsletter began to circulate on campus, everyone was talking about it. Everyone loved the jokes. Many students were quoting the newsletter in their conversations. Feeling encouraged, we planned on getting small businesses in town to advertise in our next issue to make a profit. However, we were called into the principal's office before gathering stories for the next issue.

"You need the school's permission to publish a newsletter," said the principal in a mellow voice. "It was inappropriate that you did it without permission. Some teachers you wrote about are unhappy, so I'm having this talk with you. Discontinue the newsletter, OK?"

The three of us had to nod. But we walked out of the principal's office feeling proud of the sensation we had caused in school. We made fun of the even-tempered principal by imitating his "Discontinue the newsletter, OK?" behind his back. Nevertheless, we gave up on the newsletter.

We looked for business ideas elsewhere. In 1977, I signed up for the High School Program of the non-profit Junior Achievement, through which I worked with other high school students to create a start-up company under the guidance of adult volunteers who knew how to run a business. I was elected vice president of sales for the company selling napkin rings.

We made a profit, but I was not very happy about it because almost all of the customers were parents or relatives of the student entrepreneurs'. They didn't really like or need our products.

I told myself that I would learn from this experience and my next company would make something everyone would want to buy. In 1978, I joined another venture with Junior Achievement. I ran for president. In my campaign speech, I said, "We must be creative to launch enticing products. Instead of begging customers to buy our products, we will see excited looks in their eyes when they eagerly purchase the products!"

As my inspirational speech won lots of votes, I was elected president.

After the election of company leaders, we held a meeting to decide what products we were going to make and sell. Around that time, Oak Ridge High School had just shortened the lunch break and many students were protesting against the new policy. The situation gave me an idea to produce T-shirts with a slogan advocating a longer lunch break on each of them.

We found a clothing factory to produce our T-shirts for us. Each of the T-shirts displayed the two words "Longer Lunch" and a picture of a dachshund to symbolize "long." At first our T-shirts were 100% cotton. But we soon found out pure cotton would shrink or discolor easily. After a few discussions, we changed the material to 50% cotton and 50% polyester despite a slight increase of cost.

Junior Achievement required us to do some hands-on work in the production of our T-shirts. Since we were unfamiliar with the factory machines, some T-shirts turned out to have fuzzy prints. About 16% of the T-shirts didn't look good enough for sale. We had to take them home to either wear them when doing yard work or use them as cleaning cloths.

As for selling the presentable T-shirts, we initially went door to door. But that worked very slowly. We only sold dozens of T-shirts in the first two weeks.

To largely increase the sales, we looked for whole-sale businesses and retail stores. We sold 100 T-shirts to a whole-sale shop and 60 to a local retailer thanks to the efforts of two girls on our sales team.

In 1979, I wrote the company's financial report, including a statement of earning, a balance sheet and a liquidation report. In the writing process I realized each of the investors would receive a return of $64.90. That meant our company was the most profitable venture in the history of our school's cooperation with Junior Achievement!

In the meantime, Junior Achievement announced that our corporation was "Company of the Year."

Before I graduated from Oak Ridge High School in June 1979, I also won the championship of the state math contest. With the awards, my title as vice president of the student committee, and my record of creating a school newsletter, I was chosen to be the "Most likely to Succeed" person in the year book.

My assimilation into American culture was now completed.

*Winning "Company of the Year" from Junior Achievement in 1979*

# Unpredictable Destiny

When I started thinking about college, on average one or two of the graduates from my high school would be accepted to Harvard University every year. I thought it would definitely be me.

But when my SAT scores came out, I lost some confidence. Although I got a perfect score on math, my English score was only 550, below the average performance of Harvard admits.

I knew the lower than expected English score probably resulted from my lack of motivation to memorize obscure words of the SAT vocabulary. I didn't see the point of memorizing words not regularly used.

Despite the setback, I still applied to Harvard, hoping my achievement in extracurricular activities would give me enough extra credit to help me get in. I wanted very much to enter Harvard, especially for the university's law and math programs being rated number one. At that time, I thought I would either major in math or go to law school after college.

I explained in my admission essay that my SAT English score was excusable because I was not a native speaker. I asked the admission officers to look at my strengths in science and leadership as well as my bi-cultural background that could contribute to cross-culture discussions. I thought it was quite a compelling essay. However, I received a letter of rejection from Harvard in April 1979. That was the first major frustration of my life.

More frustrating responses came. Stanford, Yale and Princeton put me on their waiting lists.

However, all the other universities I had applied to responded positively. I sent out 12 applications in total. In that pre-computer era, my teachers had to type every recommendation letter on a typewriter. I greatly appreciated their typing more letters for me than for anyone else.

I took a little time to choose between Columbia and UC Berkeley. I was leaning toward Columbia for its longer history and higher ranking. But my parents were concerned about the crime rate of New York City and preferred Berkeley. To show respect for my parents' opinion, I asked them to fly to America, to take a tour of Columbia University with me before making my decision.

When we arrived in New York, we realized Columbia University was in a safer neighborhood of New York City, and had high walls as well as campus police. That removed my parents' worries. At the same time, we were all enthralled by the beauty of the Columbia campus. The Roman-style architecture, the statues of Greek philosophers and the ivy-clad dorm buildings all presented a picture perfect ambiance of academia, with which I fell in love right away.

After accepting Columbia's offer, I ran into my friend Phillip back at Oak Ridge High School. He had just received an admission letter from Harvard. When I told him I wasn't accepted to Harvard, his eyes opened wide, "Really? Kai-Fu, I can't believe it! You used to beat me every math contest!"

Since then I've learned that life is full of surprises. It may not always give you what you want, but sometimes what it gives you is actually better than what you want. In retrospect, I think Columbia with its liberal and innovative style was the best university for me. Thanks to its allowing students to change majors easily, I was able to go into computer science, which later became the passion of my life.

I am still in touch with Phillip. He is now vice president of marketing at a telecommunications company. We often chat on line while being thousands of miles apart. His fun-loving nature hasn't changed through the years. One holiday season he signed his one-year-old daughter's name on the Christmas card he sent me. The card read, "Uncle Kai-Fu, my father asked me to send you this card, to wish you a Merry Christmas!"

# CHAPTER 4

# Learning to Make a Difference

In September 1979, I flew to New York City and entered Columbia University. Although I had visited the university once, it still amazed me that the campus was surrounded by bustling streets but looked utterly undisturbed with greenery everywhere. In the incredibly serene environment, I felt my thoughts were clearer than ever.

Columbia University emphasizes general education (GE). Students can wait until sophomore year to declare their majors. Under such liberal circumstances, most of the classes I took in my freshman year were in humanities. I studied art, history, music and philosophy, none of which were related to my later career, but all of which enriched my soul, sharpened my judgment, and helped me find my direction in life.

Now I truly understand why Columbia University lists music as a GE requirement while most other universities don't. I cannot overstate how much the music education I received there has benefited me. Since the music professors helped us explore the depth of each composer and encouraged us to attend live concerts in New York City, I have found nothing more spiritually nourishing than classical music.

Today I love Tchaikosky's piano concertos and Beethoven's symphonies even more than I did in college. The melancholy melodies that once appealed to my sentimental young heart now speak to me in a much deeper sense. Profoundly sad music somehow lifts my spirits by showing me the inevitable pain of life and making my worries all seem trivial. It also flows by me like a river that can wash away any repressed negative feelings I may have. Whenever work stresses me out, I turn on classical music and feel refreshed right away.

I am a scientist and an entrepreneur, but all the many science and business courses I took didn't influence me as much as "Contemporary Civilization," a philosophy class of my freshman year. It was from the professor of that class that I acquired my lifetime motto:.

*Imagine two worlds, one with you and one without you.*
*What's the difference between the two worlds?*
*Maximize that difference.*
*That's the meaning of your life.*

## Comparing Eastern and Western Philosophy

One day in a philosophy class, I raised my hand and asked the professor, "Why are we only learning Western philosophy? Can we use the same methodology to study Eastern philosophy? Wouldn't it be interesting to explore the similarities and differences between the two?"

The professor nodded and said it's a good idea. My classmates also expressed their interest in Eastern philosophy. Later Columbia University indeed opened new classes in Eastern philosophy and cross-culture studies. But the change took time. In my freshman year, I still only studied Western philosophers such as Plato, Aristotle and Nietzsche.

However, I did learn about Eastern philosophy at Columbia University when taking "Literary Chinese" in my senior year. The professor specified native-like fluency in the Chinese language as a prerequisite to this class because it covered pieces of classical Chinese literature that even native speakers of Chinese might not completely comprehend.

Some of the readings we did for the class were philosophy-related. They brought our in-class discussions into the realm of Chinese philosophy.

There are several schools of Chinese philosophy, among which Confucianism and Taoism (also known as Daoism) are the most influential. Confucianism in particular has dominated Chinese culture for millennia.

Confucianism began with the teachings of Confucius (551-479 BC), a thinker who recommended rigorous self discipline for the individual, a stable hierarchal structure for the family, and unchallengeable but benign authority for the government. Confucius traveled a great deal because China was divided as small countries in his lifetime and he attempted to persuade at least one of the rulers to adopt his ideas. However, none of the rulers took his advice to heart. Feeling disappointed, Confucius returned to his hometown to concentrate on teaching.

His students took notes of his lectures and compiled them into a book, titled *The Analects* (also phonetically translated as *Lunyu*), which centuries later became a must-read to all Chinese students.

It was a Han Dynasty emperor known as Han Wudi (156-87 BC) who turned *The Analects* into the Bible of the Chinese. Wudi considered the teachings of Confucianism helpful in stabilizing society and ensuring his ruling status. He mandated all teachers to disregard other schools of philosophy and teach Confucianism only.

Confucianism contains more than the teachings of Confucius. The teachings of another educator, Mengzi (372-289 BC), are also part of Confucianism because Mengzi regarded Confucius as a mentor he was born too late to meet. Like Mengzi, other scholars in later centuries expanded on the ideas of *The Analects*. All their works are considered components of Confucianism.

Simply put, Confucianism gives everyone a fixed role in society and requires all the roles to be played properly. Children must obey their parents, but parents must set good examples for their children. Likewise, a ruler must be kind enough to his subjects to deserve their loyalty.

Through the history of China's monarchy, which ended in 1911, most government officials lived by Confucianism and devoted themselves whole-heartedly to the emperor they served. But the emperor tended to disappoint them by abusing his absolute power. The officials remained loyal, following the teachings of Confucianism they had been raised with, but they needed an emotional outlet, which they saw in Taoism.

Unlike the hierarchy-oriented Confucianism, Taoism provides an egalitarian point of view. The founder of Taoism, Lao Tzu (birth and death years unknown but probably between 6th and 4th centuries BC), claims that all human beings are equal to one another and to other creatures in his book, *Tao Te Ching* (also phonetically translated more closely to the original title as *Dao De Jing*). He says, "Heaven and Earth are ruthless, with all creatures at their mercy like pigs and dogs at ours."

Lao Tzu analyzes the nature of everything, and teaches people to act like water, going around obstacles to avoid head-on conflicts. He advises against futile struggle.

Many traditional Chinese intellectuals found comfort in Taoism but considered it too passive for improving the world, so they still went by Confucianism in their careers, only applying Taoism to their personal lives. My father was one of them. He worked hard like a Confucian and minimized material desires in a Taoist way.

Taoists look upon wealth and fame as nothing to be desired because everything is transient in this ever-changing world. Besides Lao Tzu, Chuang Tzu (also phonetically translated as Zhuang Zi, 369-286 BC) also represents Taoism. He advocates a relaxed attitude towards life.

In his view, there is no point of exhausting oneself to pursue anything, including knowledge. He says, "Our lifetime is finite while knowledge is infinite. If you attempt to capture the infinite into the finite, it is bound to fail."

These two lines aroused a fierce debate in my Literary Chinese class at Columbia University.

It was interesting to study Eastern philosophy with Western methodology. I found the biggest difference between Eastern and Western philosophy to be the emphasis of the former on going along with nature and the focus of the latter on conquering nature.

The philosophy lessons I took at Columbia University deepened my understanding of both Eastern and Western cultures. Later, when I worked in China, I appreciated the understanding even more, because I saw how much Eastern philosophy (in particular Confucianism) pervaded the government system and the society. Its different way to look at the world has caused much misunderstanding between China and Americans.

*Me in New York when attending Columbia University*

# Living Happily in Poverty

Columbia University cost about $10,000 a year when I went there. I had an annual scholarship of $2,500, a student loan of $2,000 and $2,500 from my father. I needed to make up the difference of $3,000 by working part-time. I first did tutoring, and later worked in the computer center on campus.

My roommate Russ had a similar situation. Russ came from a Polish immigrant family. His father worked as a security guard in prison. His mother was a housewife. They couldn't afford his college education, so he had to support himself. He worked as a kitchen assistant in one of the university's cafeterias. Sometimes he brought leftover bread and hot dogs back to our dorm room at night. I got to share the midnight snacks with him.

I still remember the day I first met Russ in our dorm room. He was a brown-haired, blue-eyed young man about 5'10''. He smiled at me and introduced himself. Soon we became best friends.

Russ had a great sense of humor. We made fun of each other all the time.

"How come you're still not done with the programming homework?" I often shook my head in front of him and said in an exaggerated tone. "You're slower than a cow!"

I thought he might fight back with something like "not as slow as your swimming" because I had broken the slowest record in the swimming class. I didn't get any of my PE teacher mother's athletic genes.

Instead, Russ said, "Come on, buddy! You are slower with girls. You turn all red whenever talking to a girl. How are you ever going to get a girlfriend?"

That was indeed truthful. I was bold about almost everything but incredibly shy in front of girls my age. I couldn't and still can't explain my psychology in that stage of my life. I just didn't know what to say when meeting girls.

However, it didn't bother me too much because I was too busy with school and my part-time job to think about getting a girlfriend. Russ, on the other hand, often faced quite a different problem when he was unable to submit his homework assignments on time. Sometimes he asked me for help.

One day, I knew he needed me to help him with his homework, so I intentionally stayed out. Unable to find me, he went to the computer lab by himself. As soon as he logged in, he saw a warning message, "The computer will automatically shut down at 11 p.m. for maintenance." That meant he would have to complete his work within three hours, which seemed too challenging to him. He began to sweat, trying very hard to speed up his programming. When he was almost done, a pop-up window suddenly appeared on the screen to display, "Disk damaged, file lost." He was terrified, and immediately started to redo his work. But only in a few minutes, another pop-up window told him, "System damaged, all files lost. Please click on the following link."

Russ followed the instructions and saw a message:

You are fooled, Dummy! All those warnings were made up by me. I've already done your homework for you. It's in your drawer. Just come back!

*Kai-Fu*

Then Russ knew I had stolen his computer lab password to pull this trick on him. He rushed back to our dorm room, pretending to be outraged. We ended up laughing so hard that tears came out.

When it was time to go to bed, Russ rarely used his bed. The mattress was too soft for him because he had a back problem. He had taken off a closet door and placed it on the floor to sleep on it..

Every time I saw him frown, I knew his back was aching and would stop all my crazy jokes. But that only happened once in a long while. Most of the time Russ was a happy fellow. We often stayed up chatting until after midnight.

Sometimes we got hungry at night and went out to eat the cheapest fried chicken. Or we took the subway to Chinatown for low-priced midnight snacks. Once we went to a Chinese restaurant at 2 a.m. and ordered seven large plates of food, including fried rice, *chow-mien* (fried noodles) and *chow-fun* (fried rice sticks). When we asked the waitress for the check, she was shocked to see all the plates empty. She asked, "Did you guys eat up all your orders?"

We nodded.

"My God!" she uttered her astonishment. "Don't you need an ambulance?"

Like many other young men, Russ and I had huge appetites. It was often hard for us to get enough food with little money. But one winter we came up with a brilliant idea to solve the problem.

We were both staying on campus for Christmas because we couldn't afford the travel expenses of going home. In order to minimize our food budget for the winter break, Russ took 55 pounds of cream cheese from the cafeteria where he worked. We planned to make 20 cheesecakes and replace our daily meals with them throughout the winter break.

The plan worked for a few days. But then we got so sick of the cheesecakes that we couldn't even bear hearing their name. A week later, Russ suddenly announced, "Kai-Fu, I have great news! All the rest of the cheesecakes got molded. We can't eat them anymore!"

I was excited to hear that. Then we took the subway to a Chinatown restaurant that offered the lowest prices for the largest servings. We ordered six dishes to celebrate the end of our cake eating days.

From then on, we often used the word "cake" as our exclusive metaphor. While the idiom "a piece of cake" normally means "easy," we would describe the difficulty of a task by saying, "Oh! It took as much time as eating a piece of cake!" Other people would look confused upon overhearing something like that in our conversation. Then we would give each other five and laugh.

Russ kept cake making as a hobby after we graduated. Even after he moved to Germany, opened a gallery and married there, he would still send me a cake he made every holiday season. The cake sent from Germany in December would arrive in late January or February. No one in my family dared to eat it. I emailed Russ to thank him and ask him not to send any more cakes. But he replied, "For old time's sake, I must do it."

In 2000, I transferred from Microsoft Research China back to the Seattle headquarters. I was too busy moving that I forgot to notify Russ of my address change. That December he sent a cake to my old address in China. It took more than two months to get there and nearly three months to be returned. After Russ received the returned cake, he emailed me, "I always thought adding rum and chocolate was an outdated way of preserving a cake. But when I got the returned cake in May, I finally had a chance to test the old method. Now I'm happy to tell you I ate the cake, and the better news is, I'm still alive."

I burst laughing in front of my computer. All of a sudden I realized how blissful our college years had been despite our tight budgets. After going through lots of complicated life experiences, we would probably never regain the kind of simple joy that had once filled our young hearts.

In my reply to Russ, I wrote, "Glad to know you tested a five-month-old cake. Did you know that I actually mailed you a slice of the cheesecake we made in 1980? I sent it through your Polish post office. You may receive it in the next few months. Then please tell me how the 20-year-old cake tastes!"

# Changing Directions

My sophomore year at Columbia University was a turning point of my life. Before then, I always thought I would pursue a career in law or math.

I kept excelling in math throughout high school, but I was more keen on fighting for justice, so I decided to do pre-law in college and declared Political Science as my major. However, I soon found out none of the political science classes interested me in the first semester of my sophomore year.

I thought about falling back on math, but I was afraid I wasn't a real math genius. Given my record of math contest championships, I was placed in an advanced math class of only seven students. The class made me realize having been a math contest champion in Tennessee didn't mean much in front of math contest champions from states with higher academic levels such as New York and California. I also noticed my six classmates all enjoyed math more than I did. They often said they loved the beauty of math, which I didn't really see. Pure theories never appealed to me. I wanted to do something that could benefit real people in the real world.

One day it dawned on me that the part-time job I was doing in the university's computer center had helped quite a number of people. I was always able to solve other students' computer problems for them. I even wrote a program for a diamond factory's president, who had asked the computer center for programming a device that could send the number of each diamond's weight indicated on an electronic scale directly to a computer in order to eliminate the need of having workers enter numbers manually and prevent them from stealing. My quick completion of the seemingly difficult task became big news on campus.

I discovered I was far more gifted in computer programming than in math when taking computer classes. I was able to finish all the in-class assignments while my classmates were still drawing flowcharts.

I turned in my tests by the time they were half-way through theirs. They all called me a computer genius.

Their compliments reminded me of a saying I often heard, "You love what you are good at, and you are good at what you love." I absolutely loved working with computers!

That was before IBM launched personal computers. Columbia students did programming on two gigantic machines. One of them was a mainframe called IBM S/360, which was worth millions of dollars but only had a speed of 16 MHz, much slower and much more expensive than today's PCs with a speed of at least 2,000 MHz and within the price range of a few hundred dollars.

Working with IBM S/360 entailed creating, editing and storing programs on punch cards. The practice was nearly universal with IBM computers in the era. A punch card was a flexible write-once medium that encoded, most commonly, 80 characters of data. Decks of cards formed programs and collections of data. We created cards using a desk-sized keypunch with a typewriter-like keyboard. A typing error generally necessitated re-punching an entire card.

The other computer we used was DEC VAX 11/780, which was called a mini computer but actually huge by today's standards. We all loved it because it didn't require punch cards, also because its time sharing technology allowed dozens of us to use it at the same time through different terminals connected to it.

I enjoyed working with the DEC VAX so much that I couldn't wait to go to the computer lab in the evening. Sometimes I stayed there all night, and the lack of sleep made me skip classes, especially political science lectures, the next day.

I didn't want to take any more political science classes. That meant I was going to give up the option of law school. The Law School of Columbia University was rated number three at the time and would most likely promise a prominent career. By contrast, the Computer Science Department was new and its future uncertain. There were no such jobs as software engineers back then. I had no idea what I would be able to do with a computer science degree.

However, I recalled a famous quote of the distinguished journalist Whit Hobbs, "Success is waking up in the morning, whoever you are, wherever you are, however old or young, and bouncing out of bed because there's something out there you love to do, that you believe in, that you're good at -- something that's bigger than you are, and you can hardly wait to get at it again today."

That described how I felt about computer science, so I somehow had a hunch that it might lead me to success. My adventurous spirit was urging me to explore the new field as a pioneer. In the meantime, I couldn't stop pondering about computer-related questions such as "Will computers be able to think in the future? Will computers replace human brains someday?" I began to see finding answers to these questions as a difference I could make, which could be the meaning of my life!

I decided to follow my heart. I talked to my adviser about changing my major. After an in-depth conversation, he helped me with the required paperwork. The flexible system of Columbia University made the transition quite easy. I wish more universities could do the same for their students.

Nowadays I always advise college students to major in something they love, and I often use my own experience as an example. Had I stayed in political science, I wouldn't have achieved as much as in computer science. I received Bs and sometimes Cs from political science classes because studying for them felt like a chore, from which I tended to get distracted. On the contrary, I absorbed knowledge from my computer science classes like a sponge. I got a perfect score on the mid-term exam of a computer science class deemed the most difficult by everyone. The professor Zvi Galil said to me, "No one has ever received a 100 from this class. You created a record."

I impressed other computer science professors as well. For a class called "Natural Speech Processing," I collaborated with a talented classmate who shared my bi-cultural background, Lincoln Hu, to create a software program that could answer questions about natural speech like an instructor. The professor of the class gave each of us an A+ for our creativity. For another class, we did a project on moving light display. We applied some theories from a distinguished scientist Rick Rashid's Ph. D. dissertation to our project, which amazed Professor John Kender in the computer graphics class.

Lincoln found his passion of life through this project. Later he became CTO of Industrial Lights & Magic, and won two Oscars for his contribution of technology to movies.

According to Lincoln, the A+ Professor Kender gave each of us was a lot of encouragement to him. It inspired him to pursue further research in computer graphics and eventually a career in virtual reality.

The project influenced me just as much, if not more. It made Professor Kender decide to introduce me to Dr. Rashid.

"You did a marvelous job, Kai-Fu," said Professor Kender. "You should directly communicate with Professor Rick Rashid of Carnegie Mellon."

"Are you sure? John," I couldn't believe my ears. "Professor Rashid is teaching Ph. D. students. I'm only a junior."

"Why would that matter?" Professor Kender assured me. "Talking to Dr. Rashid will do your research a lot of good."

I was unaware that Professor Kender was actually helping me pave my way to graduate school at Carnegie Mellon University. I did call Dr. Rashid, and he kindly provided me guidance. Since then, Dr. Rashid has been my mentor through my career.

With Dr. Rashid's recommendation, I was accepted to the Ph. D. Program in Computer Science of Carnegie Mellon University in the last semester of my senior year at Columbia, where I was number one in the Computer Science class. My graduation GPA was 3.9, after the 4.1 from the Computer Science Department pulled up the 2.9 from the Political Science Department.

# Working Hard & Playing Hard

I often look back fondly upon my four years at Columbia University, those years when my heart was filled with youthful passion. Besides computers, I was also passionate about bridge.

After getting tired of video games in my freshman year, I spent almost all my leisure time playing bridge, which I had learned to play pretty well in high school. I knew the American Contract Bridge League (ACBL) would recognize those who had earned 300 points from bridge games as Life Masters, so I made attaining that status my goal.

Based on the ACBL rules, winning a game in a bridge club would only result in 0.3 point. That meant I had to win 1,000 games in the shortest time possible.

To that end, I played about six bridge games a week. In order to win, I went to some bridge clubs located in senior centers to play with grandpas and grandmas who didn't know the game as well as I did. But later I realized that kind of winning didn't really mean much. I began to seek challenges in competitions.

Sometimes my bridge partners and I took the train to Harvard or Yale for inter-college bridge competitions. We won the Ivy League championship.

I accumulated enough points to become a Life Master at the end of my junior year.

One of my bridge partners, Alex Ornstein, later became a professional player. He won the second place in the Bermuda Bowl, the "World Cup" of bridge. He was able to make good money and played bridge every day. It looked somewhat enviable to me that he was making a career out of his hobby. When I mentioned it to another friend, he said with a smile, "Kai-Fu, hadn't you chosen the computers, you could be a professional bridge player, too!"

Although I've never had a bridge-related job, what I've learned from bridge is helpful to my career. Thanks to bridge, I know how to read people's faces and predict their next moves. These skills are essential in business negotiations.

Drawing from my own experience, I often tell college students that extracurricular activities can do them a lot of good, and that studying is not everything. I also encourage them to get jobs or internships in the summer.

I worked two of the three summers in my college years. The first summer I obtained a great opportunity to work for the Law School of Columbia University. The dean of the Law School wanted to move a software system from an IBM mainframe to a lower-priced DEC VAX, but all the price quotes from contractors looked too high to him, so he gave the job to me based on my reputation in the university's computer center, of which he had heard.

He offered me $7 per hour, which was a high wage to me. I was excited. When he asked me how soon I could deliver initial results, I said confidently, "I can make the program run in early August so we'll have time to adjust it before school starts."

"That sounds great!" The dean looked very happy.

He believed in my promise, which I thought I could easily keep, too. But just because I considered it a piece of cake, I didn't start it from the beginning of the summer break. I obsessively played three weeks of bridge in July. Then I recalled my promise to the dean and picked up the task, which unfortunately turned out to be a lot more time-consuming than I had thought. August soon arrived. I had no choice but to try explaining to the dean, "This job is more complicated than I thought, so the program won't run until late August. But it'll still be done before school starts."

I expected the dean to accept my excuse and let me continue with the project. But he didn't. He appeared angry and said in a serious tone,

"Since you can't finish it on time, I'll get someone else to do it."

Obviously I had lost his trust. I felt terrible about it. I couldn't fall asleep that night and reflected upon myself all night long. The next day I went to the dean and apologized, "I'm sorry that I disappointed you. I broke my promise, so I'm here to return the money you've paid me."

"It's OK. You can keep the money," the dean kindly responded. "It's excusable to make a mistake when you don't have much experience. I'm sure you've learned a lesson."

I did. Since then I have always kept my promises.

This work ethic began to win applause for me during my second summer break, in the beginning of which I went to an interview for a Goldman Sachs internship. To protect its assets, the bank needed to ensure the intern's integrity, so it used a lie detector to conduct interviews.

"Do you drink?" The interviewer began with a question very easy for me to answer.

"No," I replied, feeling certain that the lie detector wouldn't indicate anything suspicious.

"Do you do drugs?"

"No."

"Have you embezzled any amount of money?"

"No."

"Do you gamble?"

"No."

"Are you sure?" The interviewer asked again. "How come your heart is beating faster?"

The lie detector! It did pick up my quicker heartbeat while I was wondering whether bridge should count as gambling. Occasionally, we'd play for money, but it was never more than $20 in one evening. "Why is your heart beating so fast?" The interviewer started interrogating me. "Do you lose money over gambling? How much do you lose every week? A thousand? Five hundred?"

He sounded more and more serious. I was afraid he already mistook me for a gambler. Would I lose this wonderful internship just because of playing bridge? I felt frustrated, but I didn't give up. I made an effort to explain my love for bridge to him, and fortunately, the misunderstanding was soon cleared.

The interviewer smiled and said, "Don't worry. You did fine. You scored a lot higher than most of the interns we've accepted."

I will never forget the two summer jobs I did in college. Before I entered the real job market, they showed me what qualities of me would be in demand, how the world would view me and how I could adjust myself to fit in better.

Nowadays, whenever a college student asks me how to make the best use of a summer break, I always say, "Get a job! Find an internship! Do something that will prepare you for the real world!"

## Meeting the Love of My Life

I didn't work in the last summer break of my college years. I went home instead. When I landed in Taipei in June 1982, I had no idea that my family was arranging blind dates for me.

My mother was concerned about my shyness in front of girls. She asked my sisters to look for girls near my age and create opportunities for me to meet them. Before I arrived home, they had already made a list of names based on the girls they had found.

I felt a little uncomfortable about meeting these girls but obliged, knowing my sisters had put in a lot of effort.

The first girl I met looked somewhat absent-minded. That made it even harder for me, shy to begin with, to start a conversation with her. While I was thinking very hard to figure out what to say, she spoke up first, "To tell you the truth, I already have a boyfriend. My parents don't like him. That's why they made me accept your sister's arrangement. But my heart really doesn't have room for anybody else. I'm sorry!"

That was the end of my first blind date. After the awkward experience, I felt reluctant about the next one. But the saving grace about the second one was, I wouldn't have to face the girl alone. It would be a lunch with many people, because my father and her father were colleagues.

The two fathers, coincidentally both from Sichuan Province, never thought about getting their children together until one of their common friends, whom I called Uncle Feng, suggested it to my father. Uncle Feng said, "Why don't you get Kai-Fu to meet Hsieh's daughter? She's a very nice girl, and pretty, too!"

My father was persuaded. He asked Uncle Feng to organize a lunch with acquaintances  from Sichuan. In the restaurant, the fathers talked about politics in their Sichuan dialect. No one mentioned a word about the purpose of the gathering. They all wanted to create a casual atmosphere that would put me and the girl from the Hsieh family at ease.

However, I was still nervous. I looked at the girl named Shen-Ling, who was sitting right across the table from me. She had wavy long hair, a cute baby face and ladylike demeanor. What a doll! I thought. But I quickly lowered my eyes to avoid staring at her, and I didn't

know what to say to her. I was afraid of saying something wrong and giving her a negative impression. She didn't say a word, either.
That night back at home, my father asked me, "What do you think about Shen-Ling?"

I hesitated. If I said I found her attractive, what if she didn't feel the same way about me? After pausing for a minute, I simply said, "I'm not sure. She was very quiet."

My father described my brief expression to Uncle Feng, who expanded on it when talking to Shen-Ling's father a few days later.

"Lee's son really enjoyed meeting your daughter," said Uncle Feng. "He could tell she's a traditional girl from her being quiet. He said she's definitely his type!"

Uncle Feng also told my father that Shen-Ling was impressed with my Ivy-League education after she actually complained to her father about my silence at the table and assumed I was probably too proud of being an Ivy-League student to talk to her.

Thanks to the diplomatic messenger, Shen-Ling and I agreed to meet again.

This time we went out by ourselves, and we talked. She was very soft spoken. All her facial expressions and gestures looked gentle. Being with her was like breathing the air of spring!

After my first date with her, I was unable to fall asleep that night. I stayed awake to recall every minute I had spent with her. The next morning I announced to my sisters, "Please don't arrange any more blind dates for me. I've found the one!"

I asked Shen-Ling out every day for the rest of the summer. I brought her a dozen roses every time I picked her up from home. The beautiful roses often caught the attention of her neighbors. They would make comments to her in front of me, saying "Your boyfriend is so romantic!" or "You guys look great together!" Then she would blush, smile and lower her eyes. I'll never forget how lovely she looked at those moments.

*Dating Shen-Ling (left) in 1982*

Shen-Ling lived in the suburbs and was unfamiliar with downtown Taipei, so I decided to take her to all of Taipei's good restaurants. My sisters liked the idea. They even chipped in to establish a "dating fund" for me so I could afford buying Shen-Ling a meal at any restaurant.

I also took Shen-Ling to the movies, the famous Shih-Lin Night Market and various snack shops. We frequently went to an ice cream store that claimed to serve more than 60 flavors. By the end of the summer, we had tasted almost all those flavors!

It was difficult to say good-bye when I had to go back to school. While flying to New York, I left my heart in Taipei.

By this time I had learned a lot about Shen-Ling. I knew she got up very early every day to do housework so her elderly grandmother and frail mother could relax. When her father was hospitalized, she stayed in the hospital for a month to take care of him. I had never met anyone my age who was taking so many family responsibilities. I admired her for being so giving to her family while most of our college-age peers only knew to take from their parents.

As soon as I arrived in New York, I wrote to Shen-Ling. There was no emailing then. I sent her a letter by air mail almost every day. She wrote back as frequently as she received my letters. But her letters were less passionate than mine. Her way of expression was very subtle.

I knew it was typical of the traditional Chinese, especially well-brought-up ladies. However, I wanted to feel more passion from her, so I came up with an idea.

I photocopied her letters, cut words out of the copies, and pasted them on a sheet of paper to compose a new letter:

*Kai-Fu, after you went back to America, I didn't sleep for three nights. I kept looking at the moon and thinking about you. I cannot get used to the days without you. I am sad, depressed and suffering from heartaches. I don't feel like doing anything. Sometimes I even feel suicidal. I've cried so much that I'm out of tears. I can't live without you! You are so smart, adorable, gentle, considerate, and perfect!*

I sent the "letter" to her and asked her to imitate the style. It fell on deaf ears.

In 1983, I was about to start a Ph. D. program. As I moved on to the next stage of my life, I wanted a special someone to accompany me, to be by my side and on my side all the time as I faced choices and challenges of life. I didn't know whether I was going to succeed, but I was sure I could make her happy.

I expressed my desire for marriage in my letters to her. She was stunned. It just seemed  too quick to her. She replied that she felt unsure about leaving Taiwan at this point because she might need to stay longer to take care of her grandmother. She asked me to give her time to think it over.

I waited a couple of weeks. Then I could no longer wait. I made an international phone call to her and said, "I know I proposed too fast, and we are still very young. But I am sure you are the one to me, and I believe you feel the same way, too. So," I paused a few seconds before formally popping the question. "Will you marry me, to make me the happiest man in the world?"

There was only silence first. Then I heard her softest voice, "I will." On Aug. 6, 1983, we held a nice wedding in Taipei.

*Shen-Ling (right), me (left) and my parents (center) at our wedding banquet*

Today's young people may find it unbelievable that I've only fallen in love once and married my first love. They find it even more unbelievable that an "arranged marriage" could work. It's also surprising to them that I married when I was only 21. But the choice I made at that immature age turned out to be absolutely wise. My stable marriage has anchored my heart and enabled me to concentrate on scientific research.

For more than a quarter century, my wife has given me the strongest support. She did all the housework when I was too busy to share the load. She never complained about being neglected when I worked overtime. She comforted me when I met obstacles in my career path. I can't thank her enough for her selfless devotion to me and our children. She gets up at six o'clock every morning in order to squeeze fresh fruit juice for our breakfast. She sews for us and irons all our clothes. While we may not hear a verbal expression of her love, we feel it in every little thing she does.

After being together for so long, I still ardently look forward to spending all my years to come with her.

*Youthful Shen-Ling and me at Yosemite National Park*

*A recent image of Shen-ling (left) and me*

**CHAPTER 5**

# Getting Recognized for "Speech Recognition"

When I started thinking about computer science graduate school in my senior year of college, there were three schools tied for number one: MIT, Stanford, and Carnegie Mellon. In the end I chose Carnegie Mellon after a free trip to Pittsburgh and a tour of the university campus.

In April 1983, Carnegie Mellon flew me there to show me the university's Ph. D. Program in Computer Science. Thanks to my communication with Professor Rick Rashid, I was considered an ideal candidate worth recruiting.

I felt flattered, knowing the Computer Science Program of Carnegie Mellon was top-rated, even though the rest of the university was not ranked so high. It was the first computer science department in the United States when founded in 1965. Since then it has always been rated number one, though currently it shares the status with MIT and Stanford.

The legend was created by the three founders of the program: Herbert Simon, a Nobel Prize winner; Alan Perlis, a Turing Award winner; Alan Newell, also a Turing Award winner.

When computer science became a new field in the 1950s, the three professors conducted research in it while teaching in the university's Business School, Math Department and Psychology Department, respectively. After founding the Computer Science Department, they looked for talents everywhere. The distinguished scientists they recruited included: Nico Habermann, an algorithm master who later became the first department chair; Raj Reddy, a Turing Award winner who specializes in speech and robotics; Manuel Blum, a Turing Award winner who masters the studies of secret codes; Ivan Sutherland, a

Turing Award winner who invented computer graphics; Dana Scott, a Turing Award winner with expertise in computer theories.

Such a marvelous program, and I was invited to become part of it --- I couldn't believe how lucky I was!

On my tour of the Carnegie Mellon campus, a Ph. D. student Joshua Bloch showed me around. Joshua later joined Sun and wrote a Java book that became recognized as the Bible of Java. Currently he is working for Google.

When I followed Joshua to visit the Computer Science building, he pointed at a vending machine and said, "This is connected with the internal network of our school so that you can check from any computer on campus to see if your favorite drink is available. Those guys (Ph. D. students in Computer Science) are lazy. They want to know what drinks and snacks are available in the machine when they are in their dorm rooms. That's why they spent two weeks figuring out how to install a chip in the machine. This way they won't come here for nothing."

What a cute thing they did! I thought, loving their idea of applying computer technology to creating conveniences in daily life.

What attracted me more was the program's unique system of matching Ph. D. students with professors. It was called a "marriage process." New Ph. D. students would enjoy lectures by all of the professors for a month, as though the professors were "courting" the students. In the end, the students would submit three of their favorites to the department. Joshua told me that most of the students would get their most favorite, and if not, at least one of the other two favorites. He said this was meant to put students in the area they were the most interested in so they would be motivated to do their best.

After listening to Joshua's description of the program, I knew it would provide me the most desirable research environment. There was no need of taking more time to think it over. I decided to accept the offer.

# Entering Carnegie Mellon

In August 1983, I flew back to New York alone after getting married in Taipei, and then moved from New York to Pittsburgh. My newly wedded wife was going to join me in Pittsburgh after I settled in.

I rented an apartment for $450 a month. My scholarship was a monthly payment of $700. That meant I would have to support myself and my wife with $250 per month. Seeing how tight our budget was going to be, I wondered if we could afford any furniture.

Just when I didn't know what to do with the empty apartment, I heard a knock on the door. I opened the door and saw my fourth sister and brother-in-law. He said, "We're coming to your rescue with a truck load of furniture."

The furniture came from my third, fourth, fifth sisters, and their husbands, who were all living in the United States at the time. They knew I couldn't afford the cost of furniture but wouldn't take their money, so they gave me their extra pieces of furniture instead.

Thanks to the furniture, the apartment looked like a home when my wife arrived. Of course she made it more comfortable.

We lived a very simple life in Pittsburgh, but just simply happy. I thrived in my Ph. D. studies and often brought my excitement home.

I chose to follow Professor Raj Reddy in his research on speech recognition after going on "blind dates" with numerous professors for a month.

Professor Reddy was a small-framed, bald man of Indian descent, in his 50s then. When he talked about speech recognition, he eyes always sparkled. He often referred to science fiction examples. He would skip his lunch break to discuss speech recognition with students, with a slice of pizza in his hand, taking a bite when listening to one of the students.

I was impressed by how passionate he was about speech recognition, which looked intriguing to me. As I was going through all kinds of research topics, I found some of them very profound but without a foreseeable application prospect, and others immediately applicable to consumer products but not deep enough to interest me. Speech recognition seemed to have the strengths of both kinds. It fascinated me to picture myself changing the way of communication between people and computers, so I listed Professor Reddy as my first choice. He happened to consider me a student he wanted to work with, too. We clicked like a perfect match.

After deciding on the direction of my research, I studied very hard. So did all my classmates. Our program held extremely high standards. The 30 students selected from more than 1,000 applicants had to take four qualifying exams in the first two years. The exams in systems, software, theory and artificial intelligence were so hard that only about 60% of the students passed at the first try. Those who failed were allowed to retake the exams later, but that would take time away from research, which was supposed to be the focus after the end of the second year. Eventually, those who repeatedly failed any of the four exams would not graduate.

Our program also had a so-called "Black Friday," which was the last Friday of every month. All the professors held a meeting on the Black Friday to discuss which students would be asked to leave. Since Carnegie Mellon annually spent a lot of money on each Ph. D. student, the university considered it a waste to invest in someone who was unable to earn the degree.

## Speech Recognition

"What's the purpose of doing a Ph. D.?" asked Dr. Nico Habermann, then chair of the Ph. D. Program in Computer Science at Carnegie Mellon University.

"To achieve important results in a certain field," I immediately replied without giving it much thought.

"No," he shook his head, looking very serious. "To do a Ph. D. is to choose a narrow and important field to conduct research, submit a world-class dissertation upon graduation, and become a top specialist in the field. Then anyone who talks about this field will mention your name."

"Yes," I felt deeply inspired and responded with excitement. "I will take away an outstanding dissertation that will change the world."

"You will take away much more than the dissertation," he said in a firm voice. "You will be equipped with the ability to independently pursue, analyze, and research any new problem or new field! You will be able to pursue scientific research, or for that matter, any pursuit of any knowledge. You will become a true scientist."

I deeply appreciated his thought-provoking remarks, which I have always kept in mind since then. Indeed, learning knowledge is shallow; learning the ability to analyze and solve any problem is something to be treasured for a lifetime.

Challenges in the field of speech recognition definitely sharpened my learning abilities. Professor Reddy asked me to create a speech recognition system that would understand everyone's speech. We called it "speaker-independent" speech recognition. At the time all the other speech recognition systems being developed were speaker-dependent. That meant each of them could only recognize one person's voice. A speaker-independent speech recognition system was a puzzle that seemed impossible to solve.

Professor Reddy said, "I think the expert system may be the best way to solve the speaker-independent speech recognition problem, and it's the hottest new technology. I hope you give it a try. Go ahead, young man! You can go all out. I have sufficient funding, so you don't have to worry about money."

I spent months developing an expert system, and achieved 91% success rate in speech recognition for speaker-dependent speech recognition. I published a paper on it and received applause. Professor Reddy was thrilled.

While Professor Reddy became more confident about the expert system, I began to have doubts because the machines run by the system were only able to understand 20 specific speakers after a very long training. I then tried to present the system with speakers it had never heard before. Given the immense variety of people's voices, replacing the 20 speakers with 100 different people would incredibly lower the success rate of the system to only 30%. Another concern was our use of limited vocabulary, only 26 words. I was afraid increasing vocabulary might collapse the system.

Summer 1984 was around the corner, but I was still at a crossroad in my research, not sure how to proceed. Then a more experienced Ph. D. student Peter Brown made a suggestion to me, "Kai-Fu, I know you are doing speech recognition and find the expert system inadequate. Why don't you try statistics? I believe drawing data from statistics can increase the success rate of speech recognition. What do you think?" Would I be able to use a large database to conduct statistics on voices? I was very curious. I felt like giving it a try.

## Mr. Lee's Hypnosis

Summer came before I had a chance to try changing my approach on recognition. I took a summer job teaching computer programming to 60 gifted high school students from Pennsylvania. This was an annual program called Pennsylvania Governor's School for the Sciences. The six-week course paid $3,000, which definitely sounded fantastic to me with a merely $700-per-month scholarship. I also considered it wonderful to gain some teaching experience and get involved in something other than speech recognition.

I enjoyed my summer class. Every morning I showed up at 8 a.m. sharp and began lecturing. I often wrote so much on the board that I used up all the chalk. I prepared lots of notes for each session, and designed all kinds of assignments for my students. I was eager to help them complete the entire junior-year-of-college course within the six weeks.

I divided the 60 students into eight groups, each to develop an algorithm to play a board game called Othello. Then they played the Othello game in pairs until one team won the championship. At the end of the course, every student had learned how to write computer programs. Their learning results were presented to governor of Pennsylvania and won his compliments. I naturally took credit for it, with a sense of achievement.

I went to the program chair to pick up my $3,000 paycheck. Then I saw a thick pile of reports on his desk. I casually asked, "What's in that thick pile, boss?"

"Oh," he replied in a low voice. "It's just your student evaluations."

"May I take a look at them?" I asked with curiosity. I definitely wanted to know what those students said about me.

The program chair suddenly looked embarrassed. He said, "It's better if you don't."

"Why? I want to improve so I can do a better job next year," I noticed something unusual in his eyes, which made me want to read those reports even more.

"You really want to teach again next year?" His response hinted a nasty surprise ahead.

I asked him again for the student evaluations. Because of my persistence, he handed me the reports, which shocked me.

On a scale of 1 to 5, I only received 1.5!

One student wrote in his comments, "Mr. Kai-Fu Lee's class was boring. Perhaps the content wasn't bad, but his interpretation was hard for us to swallow."

Another complained, "He never looked at us. He was performing a monologue all the time, so we called his class Kai-Fu Theater."

What embarrassed me most was the following description, "Mr. Lee had a monotonous tone,  which was strongly hypnotic. No matter how much I had slept the night before, I couldn't resist his hypnosis."

I turned red while reading through such comments. I didn't know my communication skills were so bad! I shouldn't have taken credit for their learning how to write programs. They were geniuses to begin with, so their achievement was not because of my good teaching, but in spite of my poor teaching.

I thought those who slept in my class were tired. How did I actually hypnotize them?

Like most other Ph. D. students, I saw becoming a professor as one of my career options. But obviously I was bad at teaching. Should I give up this option?

As I was wondering whether I would be able to improve my teaching, I thought of a Pericles quote, "Those who can think, but cannot express what they think, place themselves at the level of those who cannot think."

I couldn't allow myself to be placed at the level of those who cannot think, so I was determined to make myself a better speaker. I asked my professors for help, and they did give me pointers, including:

- Don't make a speech that doesn't even interest yourself.

- Practice three times before giving a speech. Video tape your practice.

- Make eye contact with each audience member for three to five seconds.

- If you are afraid, look at the heads of audience members in the last row. They are sitting so far that they are not sure if you are looking into their eyes or not.

I took their advice to heart and began to grab every opportunity to practice public speaking, hoping practice would make a difference, if not perfect.

## Beating a World Champion Team Member

Nowadays I give at least 25 speeches a year to more than 100,000 students. The shy, introverted young man who hypnotized 60 students no longer exists. Looking back, I feel grateful about that embarrassing experience, from which I learned to challenge myself.
Beating a World Champion Team Member

After "Kai-Fu Theater" ended in August 1984, I began to take an interest in making computers play board games.

Othello is a simplified version of the board game Go. The board is a square with eight rows and eight columns. The game begins with four discs in the center, two black and two white. Each person takes turns making moves. Each move flips the opponent's discs "captured" by the move.

To write an Othello program, it entails teaching the computer to play out possible moves and evaluate the resulting positions, in order to determine which move is best. It was once considered excellent to predict five or six steps.

The number one student of the class I had taught, Sanjoy Mahajan, designed many different algorithms for Othello, enabling the computer to predict up to seven future moves. I decided to work with him on developing an even better Othello program.

That year Sanjoy was only 16. I was 23. We devoted all our youthful passion to Othello. I took him home for dinner every night, and after dinner we went back to school to continue our research. We applied statistics to our Othello program. It was more difficult than we had imagined, but more successful than we had expected, too. In the end, our Othello program was able to predict 14 future moves.

The way we evaluated each position was using statistics based on creating a probabilistic classifier that classified each position into one of two categories: "win" and "lose." We taught the computer to generate 2,400,000 such positions, from which the evaluation function learned the chance of winning or losing.

I told Professor Reddy about the Othello program, feeling apologetic about being distracted from my research in speech recognition. But he didn't blame me at all. Instead he encouraged me and Sanjoy to enter the global competition of Othello programs.

To sign up for the competition, we needed to name our program. I suggested, "The name Othello comes from Shakespeare. Why don't we call it Shakespeare?"

"That sounds too stiff," said Sanjoy. "How about his first name William?"

"If you want to make it fun," I said. "Let's call it Bill."

For Bill, Professor Reddy financially supported me and Sanjoy to publish a research paper, lent us the best computer of the department, and flew us to California, where the world competition of Othello programs was held. He had high hopes for us, and we didn't disappoint him. Bill won all of the eight games and became the world champion of Othello programs. We were ecstatic!

After the computer competition, we wanted Bill to challenge a human world champion. A top Othello player, Brian Rose, was on the world champion team that year (he would later become an individual world champion). He was interested in finding out about Bill, too.

We decided to play three games and make the one winning two the winner.

We played the game by phone. Brian told us what moves he was going to make, and we recorded them. It was a close game in the first 15 moves. But after Brian made a fatal mistake in his 16th move, Bill's chance of winning increased to 95%. At this point, Brian was still unaware of his inevitable loss. It wasn't until a few moves later that we heard him sigh. Finally, he collapsed and declined to play further. Our Bill beat the human world champion team member, 56 to 8!

That was the first time a machine beat a human champion team member. It was a historic milestone. People became more confident in artificial intelligence than ever. Sanjoy and I published articles in *Artificial Intelligence*. He was the first high school student who published a paper in the highly esteemed publication. Later he earned a Ph. D. from Cambridge University and became a distinguished professor.

When Bill gained overnight fame, Professor Hans Berliner in my Ph. D. program told me that he was developing hardware for Chess in the hope of beating the human world champion. Given my success in applying statistics to Othello, he wanted me to do the same for Chess. He asked me to consider transferring to his team. But I couldn't leave Professor Reddy after he had done so much for me. Also, I didn't want my dissertation to be all about board games. I was hoping to make something more useful.

## The Most Important Scientific Innovation of 1988

Professor Reddy received a $3,000,000 fund from the Department of Defense to conduct research in speaker-independent, large vocabulary, continuous speech recognition. That meant the objective was to make computers understand everyone's pronunciation, thousands of words and the continuous flows of words. Professor Reddy recruited more than 30 professors, researchers, phoneticians, linguists, programmers and students to work on the unprecedented program. He also wanted me to cooperate with the team, to make breakthroughs in the expert system.

However, I was losing interest in the expert system. Based on my Othello experience, I believed establishing a large database would be the way to reach our three goals in speech recognition. But how was I going to bring this up to Professor Reddy? He had already received government funding for a research project on the expert system, so there was no way to drop the project.

How would he respond if I told him I would take another route to improve speech recognition? Would he be upset? Would he try to persuade me to continue working on the expert system?

I felt uncomfortable about bringing up my different view to him, but I still did it because I recalled what our program chair, Dr. Habermann, said about the purpose of doing a Ph. D. I couldn't fail Dr. Habermann's expectations. Nor could I waste my Ph. D. years on a project in which I saw no future.

Mustering all my courage, I said to Professor Reddy, "I'd like to take another approach, to use statistics to achieve speaker-independent, large-vocabulary continuous speech recognition."

To my surprise, he wasn't offended at all. He asked in a mellow voice, "How are you going to solve the three problems with statistics?"

I was prepared to answer this question, so I explained my ideas non-stop for the next 10 or 15 minutes. After listening to me, Professor Reddy calmly said in his always gentle tone, "I disagree with your point of view on the expert system and statistics. But I can support you to do it with statistics, because I believe in science, there is no absolute right or wrong, and we are all equals. I also believe someone with passion can find the best solutions."

I couldn't express how touched I felt at that moment, so I was speechless. When a student decided to go against his professor's instructions, the professor was still willing to provide strong support. This would be unimaginable in many places!

In order to build a large database for the statistics I intended to do, I needed funding. Professor Reddy came to my aid as promised. He said, "We are not a professor and a student in the field of science. We are conquerors of problems. So, if you need a database, I'll convince the Department of Defense to help you build a large one."

My statistics also required quick computers. Professor Reddy bought the latest Sun microcomputer for me. Every time our department received a new machine, he always said, "Ask Kai-Fu if he needs it."

My project cost him hundreds of thousands of dollars.

His big-heartedness made me feel a great power, a power of freedom and trust. He reminded me of a Voltaire quote, "I may disagree with what you have to say, but I shall defend to death your right to say it."

Professor Reddy demonstrated a scientist's spirit by saying, "I disagree with you, but I support you." His belief in everyone being equal in front of science deeply influenced me. I vowed to follow his leadership style.

Twenty-four years later, my former employee Alan Guo expressed his gratitude to me in the same way I felt about Professor Reddy. He said, "Kai-Fu has taught me that you can sincerely disagree and full-heartedly support at the same time. I'll remember this for the rest of my life. When I first learned about how Kai-Fu's Ph. D. adviser supported his different research project, I thought it was just generosity. But after Kai-Fu doubted but supported my decisions again and again, I've realized it's a rare kind of leadership."

I owe this leadership to my mentor, Professor Reddy.

Thanks to his support, I devoted myself to my research project like a workaholic. After doing school assignments in the morning, I went home for lunch and then worked from 1 p.m. to 2 a.m. on the project. I worked 18 hours a day, six days a week. I took a break on Sundays as promised to my wife. But even on Sundays I would take a few chances to check how my experiments were doing. From the end of 1984 to the beginning of 1987, I worked with another Ph. D. student to apply statistics to speech recognition. More than 30 other people were using the expert system to tackle the same problems. We were competing with them in terms of methodology. But we shared everything under the leadership of Professor Reddy. We used the same samples and tests.

By the end of 1986, the statistics system and the expert system reached about the same level, 40% of recognition.

In May 1987, we largely increased our training database. I came up with a new method, triphones, which could help the computer not only learn every sound but also recognize the transition between every two sounds. Since we didn't have enough samples of certain sounds, I created another method, generalized triphones, to combine other sounds into them. Then our recognition rate rose to 80%!

The success filled my heart with joy. Professor Reddy was also very happy. He decided to bring my research results to an international conference, to show the world that statistics worked to improve speech recognition. Seeing this wonderful opportunity, I asked, "May I go with you to present my research results?"

"Sure!" He immediately agreed. "That'd be great! I'll book a plane ticket for you right away."

On the day of our departure, I drove my 15-year-old Toyota Corolla to pick Professor Reddy up. Unexpectedly, we saw smoke coming out of my car when we were about halfway to the airport. The engine busted. We had to stop, get the car towed and take a taxi.

*Standing in front of the busted Toyota Corola in 1987*

Despite the delay, we got on our flight and made it to the international conference. My presentation caught a lot of attention. Several IBM fellows and Stanford professors asked me to explain further details of my project to them.

After the conference, I worked even harder, in the hope of making the system's recognition rate even higher. I stayed up until I could no longer open my eyes.

One morning, when I opened the file containing the experimental results from overnight, feeling somewhat sleepy, I suddenly noticed the recognition rate was 96%! Was I dreaming? I rubbed my eyes and looked again. Yes! It was 96%! I was so excited that I almost fainted. Oh yes! It must have been the revision of certain details I had done the night before that made the breakthrough.

That day, I stopped working, and took my wife to a nice restaurant to celebrate, because I knew that moment, I would graduate meeting Professor Habermann's expectations.

In April 1988, I was invited to attend the annual Speech Recognition Conference. About a month before the conference, Professor Reddy taught me one more important lesson.

He said, "For your 30 minutes, you should only talk 25 minutes, and then let the audience try out the system in the last five minutes."

I said, "But it'll be noisy over there. It'll affect the recognition rate, making it lower than 96%. There will also be many Japanese scholars whose accent my system has never been exposed to."

Professor Reddy explained, "It actually doesn't matter whether the recognition rate is 96% or 90% when you demonstrate it over there. You are there to make an impression. Then all those scholars will remember their first contact with a speaker-independent speech recognition system was in New York, at Kai-Fu Lee's presentation."

"I understand," I nodded, feeling grateful about the marketing strategy he had just shown me. "But I'm afraid the recognition speed is too slow. It may not look good to keep people waiting."

"That's no problem," Professor Reddy said. "I'll get Fil Alleva, our best engineer, to help you revise the program, to make it run faster."

"By the way," he added. "You should name your system so they know how to call it. Your Othello system has a name."

I named the system Sphinx, for its lion body to represent a large database, its human head to symbolize knowledge, and its bird wings to signify the speed of the system.

As predicted by Professor Reddy, the demonstration of my speech recognition system amazed all the conference attendees. It was deemed the most distinguished achievement in the field of computer science. Those speech recognition researchers working on the expert system all began to turn to statistics.

The New York Times heard of the demonstration and sent a reporter, John Markoff, to interview me in Pittsburgh. The article came out on July 6, 1988. It took half of the first page in the newspaper's science and technology section. I was impressed by how much Markoff understood my research. Later I learned that he is a talented reporter, three-time Pulitzer nominee, and part-time instructor at Stanford University.

More media coverage came my way. BusinessWeek magazine selected my speech recognition system as the most important scientific invention of 1988.

I felt very fortunate about making such a success at age 26.

## An Exceptional Job Offer

I received my Ph. D. from Carnegie Mellon University in April 1988, four and a half years after I entered the Ph. D. program. Normally it takes six years to earn a Ph. D. in computer science from Carnegie Mellon. I was among the fastest.

My whole family was excited about it. They all flew to Pittsburgh for my graduation.

The ceremony began with bagpipe music. The band members were all in Scottish kilts.

One graduate after another went onto the stage to receive the diploma from the university president. At the end of the ceremony, we all threw our caps into the air. One of my family members caught my cap

*With my father (right) at my Ph. D graduation in 1988*

and put it on his head. Another grabbed it from him and did the same. We took lots of photos.

We continued to party after returning to my place. My wife made more than a dozen delicious Chinese dishes. When everyone sat down to eat, I noticed a sense of pride in my father's eyes. That was the first time he looked at me that way. There had always been a little concern in his eyes on me before that day.

I had received quite a few job offers with decent compensation packages from high tech companies such as IBM, Apple and Bell Labs before graduation. But Professor Reddy came to me before I chose one of them. He said, "Kai-Fu, I know you have many choices in front of you. But I hope you can stay at Carnegie Mellon. Generally we don't suggest Ph. D. graduates stay here to teach, because we prefer new blood. But your achievement in speech recognition can bring funding from the Department of Defense to the university. Because of that, we can let you skip the post-doc stage and make you a research computer scientist. If you want, you can switch over to an assistant professor later, with all your time spent counted towards tenure."

"However," he added. "We pay less than those high tech companies."

I didn't immediately respond as I was thinking about this offer as well as all the other options. Professor Reddy thought the lower pay sounded discouraging to me, so he began to do some math for me.

"Kai-Fu, if you go to Microsoft, Apple or IBM, your annual salary will be about $80,000. If you stay here, the annual salary is $51,000. But," he raised his voice. "You only work four days a week here. You can use the fifth weekday to work as a consultant for those high tech companies. That'll be about 1,000 a day. You can do it 50 days a year. That's an extra income of $50,000."

"$5.1 + 5 = 10.1 > 8$," Professor Reddy showed me this simple math formula. "What do you think, Kai-Fu? Does it sound worth it to you?"

I smiled, feeling touched by his thoughtful consideration for my future earnings.

"Come, young man!" He patted my shoulder. "Join us!"

After accepting Professor Reddy's offer, I became one of the youngest faculty members at Carnegie Mellon. I bought my first house for myself and my wife to settle down in Pittsburgh.

I continued to do research in speech recognition. Professor Reddy put a portion of his funding from the Department of Defense in my project. I also annually received $60,000 from the National Science Foundation and $40,000 from Texas Equipment, for which I served as Principle Investigator. With about $200,000 as my annual budget, I recruited a post doc, Xuedong Huang, who had earned his Ph. D. from the University of Edinburgh. I also selected three Ph. D. students at Carnegie Mellon to do team work on speech recognition. The five of us kept our project number one in the Department of Defense's annual evaluations.

As predicted by Professor Reddy, I received invitations from many companies to work for them as a consultant. That enabled me to travel all over the United States. For each assignment, I would fly to the destination on the weekend before and take a tour before going to the company on Monday.

I most frequently flew to San Francisco and Los Angeles. Whenever I went to LA, I would go to Todai Buffet for the all-you-can-eat sashimi and sushi. A few years later, the restaurant grew into a national chain. But I discovered it before it became famous.

In addition to domestic trips, I also flew to Tokyo and Taipei sometimes to give presentations. I found the Japanese even more conservative and reserved than the Chinese. When I entered a conference room in Tokyo, all the researchers sitting in the room would stand up and clap their hands to welcome me. They were all wearing white uniform coats like medical doctors. During my speech, all of them were looking at me attentively. At the end of the speech, I told them it was time for them to ask questions. But everyone remained quiet. No one raised a question.

## Going Back to My Roots

While I continuously sought new ways to improve speech recognition in computers, my father devoted himself to studying the 20th-century Chinese history. He said he had witnessed a lot of that history and therefore felt responsible for leaving an accurate record of it to future generations.

In order to find uncensored materials related to the Chinese Nationalists and Communists, my father came to the United States. He stayed at my house in Pittsburgh for six months. Every morning I drove him to the East Asian Library at the University of Pittsburgh on my way to work, and picked him up after work. He spent the whole day in the library and would just buy a sandwich from the library cafeteria for lunch.

My wife and I were touched to see how hard-working he was at age 80. One day I noticed he had written a motto for himself and placed it on his desk. It read, "Knowing the sun will set soon, the old horse runs faster without being whipped."

My father could never forget his homeland. He often asked me, "Do you think maybe you could go back to China and teach?"

I didn't really understand how he felt. I had a very vague idea about China, though my family came from there. All my parents could tell me about China was what happened before they left. As for up-to-date information, I only gained a little during my Ph. D. years from a fellow Ph. D. student, Weimin Shen.

Weimin surprised everyone when he wrote programs on paper. We asked him, "How come you don't write programs on the computer?"

He said, "This is the way we do it in China. We don't have so many computers to let each student use one. That's why we write programs on paper to submit to the professor."

This seemed unbelievably difficult to me. I learned from Weimin that China had a backward research environment in computer science. But I didn't think of doing anything to improve it until I saw it with my own eyes.

In 1989, the Beijing Institute of Information Technology received a research fund from the United Nations and invited me to open a four-week course there. My plane landed in Dalian, an industrial city northeast of Beijing, due to weather conditions. When I walked out of the Dalian airport, I saw crummy buildings, dingily dressed people, narrow streets and bicycles everywhere.

I took the train to Beijing. The capital city looked better than Dalian but not by far.

The cafeteria of the Beijing Institute of Information Technology was decrepit. During lunch, every student took a big container there, put some food in it and then started eating on their way back to the dorm building. The cafeteria would prepare a few extra dishes with higher quality for me. One day I was watching the cook making my lunch, and I asked him, "Why are the students eating and walking at the same time?"

The cook shrugged, "They've always been like that. You don't do it in America?"

Although the students never sat down to have lunch with me, they took me out to have dinner at a different restaurant almost every evening. They also recommended some famous restaurants for me to try on my own.

One night I went to a restaurant named Dong-Lai-Shun based on their recommendations. The restaurant was renowned for its hot pot, also known as the Chinese fondue, which has the set-up of a fondue but no chocolate or cheese in it. Instead there is savory or spicy boiling broth in the pot for people to quickly boil thin slices of meat, seafood as well as vegetables and then dip them in a sauce before eating.

When I ordered a small hot pot at Dong-Lai-Shun, the waiter said, "We are not serving hot pot right now because we'll close soon. But we can make a plate of stir-fried chicken for you."

I was shocked. It was not even 8:30 yet. The restaurant was already closing? I shook my head, "I came here for your hot pot."

To my surprise, the waiter just wouldn't accommodate, "It's too late for hot pot. And if you don't order now, it'll be too late for stir-fried chicken, too!"

I had never seen a waiter being so tough. I almost walked out. But it would take time to walk to another restaurant, and I was hungry, so I agreed to order stir-fried chicken.

Just after I took the second bite of the stir-fried chicken, the waiter rushed to my table and said, "Please pay now, comrade. I'm getting off work."

"What?" I raised my head while being taken by surprise, and my eyeglasses almost fell off. "I'm not done eating yet. I have to pay now?"

"Yes," he continued to push me. "It's time for me to go home. I'll leave as soon as you pay."

He did hurry out of the restaurant right after I paid, leaving dirty dishes on some tables. I could hardly believe my eyes!

Later I realized that was a die-hard old habit from China's planned economy while the country was moving toward a market economy. The waiter had little incentive to provide impressive service, when his job was both secure and low-paying, no matter how good or how bad a job he did.

That was before China took off economically. I gave lectures on the most advanced technology in a backward environment. My classes were always fuller than capacity. Students from other schools came. Many of them couldn't get seats and had to stand. But the standing ones managed to take notes, too. All of them looked attentive every minute through every lecture.

I felt their thirst for knowledge. Looking at them, I thought of my identical ethnic background. Hadn't my parents left China, had I been born in the country, I could have been one of those students struggling against a difficult environment.

I saw great potential in the diligent students. It was only the lack of resources that was holding them back. With this realization, I was happy about bringing them some resources they desperately needed, and I felt like doing more.

That planted the seed of my later effort to advise Chinese college students.

# CHAPTER 6

# Leaving Academia for Apple

In 1990, I converted from a research computer scientist to an assistant professor. Being an assistant professor at Carnegie Mellon University was like living in paradise, secure and carefree. I didn't have to worry about the market or the economy like those working in the business world. My position was tenure-track, and I was valued by the department as the primary expert in speech recognition. That meant I would be able to obtain tenure in a few years and be set for life.

Surprisingly, I felt a sense of loss while having such job security. I wondered if I really wanted to stay in academia for the rest of my life. While loving research, I disliked some non-academic duties required of a professor, such as socializing with Washington funding agencies in order to get federal funding and going to all kinds of university staff meetings that had nothing to do with my expertise. Even my research sometimes frustrated me. My papers just sat there. No one was making use of them. What differentiated them from wastepaper? I felt being stuck in an ivory tower, totally disconnected from the real world where I desired to make a difference.

One day I received a phone call from a stranger. He said, "Dr. Lee, two vice presidents of Apple are interested in you. They think someone like you should come out to make some real products. Would you be willing to have a talk with them?"

This was two years after I received my first job offer from Apple. I felt absolutely flattered because Apple in 1990 was the leading computer company in America, and it was considered just as cool as it is today. People believed in Apple almost like a religion. They were more than willing to pay a much higher price to buy an Apple computer, which was artistically made and looked seamless. Every Apple Worldwide Developers Conference attracted countless fans. I was also a fan. I bought a new Mac on the first day it hit the market. As a speech scientist, I was fascinated that the Mac had a pre-installed MacinTalk, which talked to me in robotic English.

"How cool would it be if I could work for Apple!" I heard such a voice from my heart.

## The Groundbreaking Apple

It is widely known that Apple opened up the era of personal computers. In 1983, Apple earned 980 million dollars, and the 28-year-old Steve Jobs took 284 million, becoming the youngest among America's 40 richest people.

Jobs was so proud of his achievement that he underestimated his competitors. In 1984, when Mac first came out, a news reporter asked Bill Gates, "How soon will you port Mac Excel to Microsoft PC?" Gates replied, "It'll probably take some time---" But Jobs interrupted, "It won't happen even after we die."

Jobs passed his overconfidence to his employees. Every time Microsoft launched a new Windows system, Apple held a staff meeting to expose its flaws and talk about how much it was behind Apple products. However, every new version of Windows came closer to Mac. Some Apple employees began to wonder how many people would still pay double to buy Apple products when the difference was only 5%.

Apple fans didn't look at it this way. They said, "Windows and Mac may have 95% of similarities. But that's just like a transsexual woman and a real woman. They may be 95% the same. But the 5% difference is what we care about most."

Even so, Apple began to lose market share when Wintel (Windows + Intel) PCs left Apple behind the dust.

Jobs knew he needed someone to revive Apple's sales and marketing, so he went to John Sculley, who had been president of Pepsi and helped Pepsi beat Coca Cola for the first time in history. Jobs persuaded Sculley to lead Apple with a question that later became a famous quote, "Do you want to spend the rest of your life selling sugared water or do you want a chance to change the world?"

Unexpectedly, Sculley and Jobs didn't get along when working together as CEO and vice president. Jobs blamed Sculley for the continuing drop in the Macintosh sales (merely 2,500 units in March 1985). When Sculley learned that Jobs intended to remove him and take his place, he decided to take the matter to a board meeting in May 1985. He knew all the board members thought of Jobs as a inspirational motivator but not enough of a day-to-day manager to be CEO. As Sculley expected, the board members all supported him and backed up his decision to dismiss Jobs from his positions as vice president and the leader of the Macintosh Division. Jobs was so upset about it that he took long vacations after that and submitted a resignation letter in December 1985. He was basically forced out of the company he had co-founded!

Without Jobs around, Sculley became the unequivocal head of Apple. In the next few years Apple's laser printers and desktop publishing software enabled Mac to monopolize the publishing industry in America. In 1989, the Macintosh sales increased from 300,000 to 3,000,000. Apple became the hottest company on Wall Street and Sculley the highest paid manager in Silicon Valley.

Sculley kept looking for new product ideas. He recruited a psychologist, Dave Nagel, to be vice president of Apple's Advanced Technology Group. The group was interested in speech recognition. So was the Mac III Product Team, led by Hugh Martin. That was how Dave and Hugh decided to contact me in 1990.

I flew to Cupertino, California to meet them. As I arrived, it was very pleasant to see the evergreen scenery and ever-bright sunshine. Perhaps because people would naturally eat less and move more in warm weather, everyone I saw in Silicon Valley looked slimmer than those in Pittsburgh.

Apple employees welcomed me with friendly smiles. They all looked optimistic and confident. Most of them seemed very young.

The two vice presidents were older. Hugh was a middle-aged man. His eyes sparkled when talking about the prospect of Mac. Dave was older than Hugh. He had a white beard and a smile that reminded me of Santa Claus. He invited me to dinner.

"You know, Kai-Fu," said Dave. "The first GUI (graphical user interface) was developed at the Xerox PARC (Palo Alto Research Center), but it didn't result in a successful product. It was Apple that made GUI useful to many people. So you see how Apple inventions influence people."

"We are doing Mac III and would like to incorporate speech recognition in it," he added. "Kai-Fu, don't blame me for copying Steve Jobs by asking you this question: do you want to spend the rest of your life writing useless research papers, or do you want a chance to change the world?"

His clever revision of the Jobs quote directly spoke to my heart. To me, there was nothing more enticing than making a difference in the world. At that moment, my heart decided for me: I was going to join Apple!

My only concern was how to submit my resignation to Professor Reddy. I was afraid of disappointing him. Would he try to make me stay? How would I respond if he did?

I took a deep breath before going into his office.

"I'm sorry to surprise you with a decision I just made," I said slowly, using my most cautious tone. "I've thought about it for quite a while. Apple asked me to work for them to put speech recognition into their products. I consider Apple a very cool company, and I like what I've seen of their technology. So," I paused a few seconds and finally said, "I'd like to go."

"Oh?" Professor Reddy seemed to be taken by surprise. He was speechless for a minute. Then he seriously looked at me. "Have you really thought it through?" he asked.

"Yes, sir."

"That's good," he said in an understanding tone. "Not everyone should always stay in the field of research. If you think working for Apple can further develop your talent, then you should go. Do a great job!"

Every time I made a choice, Professor Reddy respected it and supported it like a loving father. Every time I felt his best wishes for me to realize my dreams. When I turned around to leave his office, I felt like crying. At this moment, I heard his voice.

"Kai-Fu, are they giving you good enough resources?"

I turned back to reply, "Yes. They'll let me join a strong team. They'll let me hire some young people. They'll let me work on the research I want to do."

My mentor nodded, with an encouraging smile.

## A Pirate Culture

I began working for Apple on a sunny day. When I looked at the directions to my workplace, I was surprised. There was a sign that read, "Cupertino National Bank." How could my office be in a bank? Why wasn't it in the Apple headquarters?

I arrived at the bank and asked the security guard, "Is there an Apple office here?"

He pointed at the back door.

I went to the back door and found a stairway behind it. I walked upstairs. When I arrived on the second floor, I saw several young people working in front of computers.

Apple had this secret office because the company wanted to keep product development absolutely confidential. Then the products would astound everyone when they came out.

"Perhaps this is a reflection of Apple's pirate spirit," I told myself, knowing there was a pirate spirit in Apple. Steve Jobs once raised a pirate flag on top of his building. He also placed a Besendorfer piano and a $10,000-worth pair of speakers in the lobby. Beside the piano was a BMW motorcycle, which guests would see right away when they entered the building. Jobs would play the piano for the employees working late at night. I imagined how passionate music must have filled the lobby at those moments and somehow wished I could have been there.

Apple managers often told employees, "Work on your inventions, don't care about what people say, and believe that one day we can change the world."

The Apple employees I worked with were all very young. I was 28 then. They were around my age or even younger. They loved Apple's pirate culture. Some of them brought their pets to work. Once a colleague's rabbit jumped on my hand while I was typing. The atmosphere was very casual.

Among my colleagues were Philip Miller, who once co-authored the famous software program Lotus 1-2-3, Phil Goldman, who created Web TV, which was sold to Microsoft in 1997 for $425 million, and Andy Rubin, who years later took charge of Google's Android.

These remarkably talented colleagues helped me transition from pure research to product development. We worked harmoniously together in the hope of making computers ultimately become people's intelligent assistants. We wanted computers to understand human language and follow verbal instructions. We were dreaming of three-dimensional user interface, which could be applied to games, social media, and video conferencing.

From July 1990 to February 1991, I had a fabulous time at Apple. No one managed this team. We had maximum freedom to maximize our potential. We were so into our work that sometimes we forgot it was time to go home. Our tireless effort made the speech recognition in Mac 40 times faster.

We also made a few breakthroughs in Mac III, including video conferencing, speech recognition, speech synthesis, video camera, three-dimensional user interface, telephony, and high-definition sound effect. The product would use the Motorola 88110 processor and a Mach operating system developed by the Computer Science Department of Carnegie Mellon University.

## A Financial Crisis

In the Oscar-winning movie "Forest Gump," the main character inadvertently makes a lot of money from Apple's stock. That segment of the 1994 movie reflects how people viewed Apple in the early 1990s.

However, Apple was in fact struggling financially in those years. The not-so-big company had more than 1,000 projects going on at the same time, but few of them turned into real products. Apple was losing market share to Microsoft.

This was partially because Apple stayed away from the industry standards set by IBM and Microsoft. Starting in 1986, IBM and Microsoft made their specifications of personal computers industry standards, so their products were compatible with one another. This resulted in "horizontal integration," where in microprocessors, motherboards, and software the standards led to higher volume and lower cost. As consumers welcomed the convenience of such compatibility plus lower prices, the IBM-Microsoft standards gradually became mainstream.

Sculley was aware of the trend and wanted Apple to take part in it. He suggested developing a Mac operating system on the PC platform. The board agreed. But the Macintosh Division vehemently opposed the idea. A software business would cannibalize their hardware sales. Also, as Apple engineers viewed their products as artworks, there was no way for them to share their unique creations with other companies for mass production. That just sounded like something beneath them.

Since Apple was unwilling to make Mac graphic user interface transplantable to PCs, Microsoft worked on creating the same technology and then produced a clone version of the Mac graphic operating system, Windows.

When Windows came out, Sculley accused Microsoft of copying Apple and asked Bill Gates for a serious talk.

"We didn't copy Apple," said Gates. "Both Apple and Microsoft learned the technology from Xerox PARC."

"But Apple was authorized by Xerox PARC to use the technology. You didn't get the authorization. We also have patents, " Sculley said.

"If you are holding this against us," Gates stopped defending Microsoft and started threatening Apple. "We'll stop all the Mac-compatible software development, including Office."

This was a real threat. Apple could not afford to lose Mac's compatibility with MS Office.

"If you promise your next-generation software will be Mac compatible, we can give you a one-time authorization to use our patents," Sculley conceded.

During the patent negotiations, Microsoft changed "one-time" to "current and future" about the authorization, and somehow Apple acquiesced. This was recorded in the documents signed by the two companies. Apple therefore lost 179 patents.

Later Apple sued Microsoft. But based on the signed documents, the judge only asked Microsoft to modify a few features (such as renaming the "trash can" as "recycle bin"). Apple was unable to gain anything back.

In short, Apple's market share kept shrinking because it insisted on maintaining a high-end image and wouldn't open any of its specifications. Sculley once tried to increase market share by lowering prices, and it did help the sales. But the company was not set up to sell volume products at lower prices. Margins shrank, and the company faced multiple quarterly losses.

In the end, Apple had to save costs through layoffs. After the first layoff I witnessed, some let-go employees came back marching in a group. But they were not protesting. Instead they were cheering for the company. They said, "Our blood is in six colors (like Apple's then-six-colored logo)!"

Following the lay-offs, Apple reorganized several groups. One day, my boss Dave Nagel came to my office and said to me, "I have good news and bad news. Which one would you like to hear first?"

I was taken by surprise. But I soon calmed myself down and said, "I'll hear the bad news first."

Dave said, "Your Mac III team was eliminated."

I was shocked and saddened.

"But the good news is," Dave continued. "Speech recognition will be kept and moved to the ATG (Advanced Technology Group). You'll be promoted as the manager of the speech recognition team in the ATG."

"What?" I was astounded. "But I don't have any management experience!"

"I believe you have potential in management because all your colleagues say they like working with you. And your new boss, Shane Robinson, is an outstanding leader. He'll teach you what you don't know."

Dave was right. Shane did give me a lot of helpful guidance later. Today Shane's leadership shines in his position as CTO of HP.

## Casper Teasing Joan Lunden

After I became the manager of the Speech Recognition, Speech Synthesis, and Natural Language Processing Team, our research delivered impressive results. John Sculley heard of our achievement and told us to run a demonstration for him.

The demonstration was scheduled for Dec. 16, 1991, which happened to be the day my eldest daughter Jennifer was born. She came into the world at 1 p.m. I left the hospital at 3 p.m. for the presentation. No one knew the product as well as I did, and no one was as good as I was in talking to the computer. I had to be the one explaining it to Sculley.

At the end of the demonstration, Sculley's eyes sparkled. He walked onto the stage and announced, "Your product is absolutely significant! It's astonishing! I'll take it to TED (the Technology, Entertainment and Design Conference). But the most exciting thing today is not this, but that Kai-Fu got a daughter. Let's congratulate him!"

My colleagues gave me a huge round of applause.

In February 1992, Sculley appointed me to speak at the TED conference, where there were almost as many celebrities as at an Academy Awards ceremony. There were high tech company CEOs and science professors as well as movie stars and directors. There were also many reporters. I was a little nervous when I walked onto the stage. But fortunately the focus of everyone's attention wasn't me, but the computer that was following my instructions, the computer with a cute cartoon character's name, Casper.

At the end of my presentation, all the audience members stood up to applaud. Many went to Sculley to shake his hand. More people came to me to ask further details of the technology. At that moment, I felt like I was in heaven...

My presentation became headline news in the Wall Street Journal the next day. Then I realized many of the people asking me technical questions at the conference were reporters.

Apple's new technology aroused a lot of curiosity in the media. In March 1992, Sculley received an invitation from Good Morning America. That meant we were going to take Casper from our high tech circle to a general audience. Sculley had me accompany him to the live show.

Before we flew to New York, I cautioned Sculley that our system was new and hadn't been perfected. It would look really bad if Casper suddenly died on the air. Then Sculley asked me, "How likely would that happen?"

I said, "About 10%."

Sculley asked, "Would it be possible to lower the failure rate to 1%?"

I thought about it for a couple of minutes, and then replied, "OK, John, I'll give it a try."

On March 12, 1992, Sculley and I took our Casper to Good Morning America at 7 a.m. We knew there would be more than 20 million people watching us while eating their breakfast.

The program host Joan Lunden introduced us. She said, "Reality is a step closer to science fiction with Apple's newly developed program that allows computers to understand and respond to spoken command. For us to take a first look at that, joining us are John Sculley, Chief Executive Officer of Apple, Kai-Fu Lee, the inventor of Apple's speech recognition technology and the computer called Casper. Nice to have the three of you here!"

Joan greeted Sculley and me. She also said, "Casper, good morning!"

"Good morning America! And good morning Joan!" Casper responded exactly the way I had trained it to do. A good start! Both Sculley and I felt relieved.

I began demonstrating how Casper worked by telling it to do some editing work in a document displayed on the computer screen and on camera. Sculley asked it to pay bills. Joan had it program her VCR to record Good Morning America and other TV shows. Casper fulfilled all the tasks by showing clicks on the computer screen.

Then Sculley gave it a more complicated assignment, "Please schedule a meeting for me with Bob Strong."

"What day would you like to meet?" Casper asked.

"Wednesday," Sculley replied.

"At what time would you like to schedule the meeting?" Casper asked.

"From 2 to 3 p.m.," Sculley said.

"Your meeting with Bob Strong is scheduled for 2 p.m. on Wednesday," Casper said.

"Wow!" Joan seemed amazed. She praised Casper's communication skills. Then she smiled and asked, "Is it going to be some software you can adapt to other computers to do such tasks as scheduling meetings? "

Sculley was about to reply, but Casper spoke up first, "What day would you like to meet?"

Joan burst laughing. Casper had been trained to react to the word "meeting." But in this context it sounded like asking Joan for a date.

After the memorable funny scene was broadcast, Apple's stock immediately rose from $60 to $63 per share.

Later, Sculley asked me, "How did you lower the failure rate to 1%?"

I replied, "I brought two computers and connected them. Had the first one gone down, the second one would have taken its place. Because a computer's failure rate is 10%. The failure rate of both is 10% X 10% =1%. So our success rate was 99%!"

Casper made Americans believe Apple was still an innovative company despite its continuous loss of market share. I felt a strong sense of achievement. Although many other companies attempted to recruit me, I was not tempted. Like most of my colleagues, I loved Apple.

On April 1, 1993, many Apple employees saw a speaker inside the elevator and a sign that read, "Tell me what floor you want to go, and I'll take you there." The speech recognition system had been applied to the elevator. Someone tried it out by saying to the speaker, "Second floor." The "2" button really lit up! Unfortunately, just like on "Good Morning, America", the system would trigger on false alarms. If someone said, "Wait a second," in a conversation, the "2" button would light up too!

Later I learned that some engineers on my team had secretly connected our system with the elevator. It was a joke for April Fools' Day. Also I found my demo machine had disappeared. On April 2nd, it was returned to me, full of dust, as it had sat on top of the elevator for a whole day.

## Learning the Art of Management

Apple products were regarded as artworks, and Apple engineers were somewhat like artists. As a manager, I believed the best way to manage these free spirits was to give them license to do what they wanted to do as long as they wouldn't cross the line when it concerned the well being of the company.

I was learning the art of management through experience at Apple. When I first became a manager there, I was only 31. Some of my subordinates were older than I was and had higher seniority than I did.

An older scientist in his late 50s didn't feel good about me being his boss. A very well known expert in his field, he often argued with me at meetings and purposefully went against my decisions. I didn't want any conflicts with subordinates, so I kept holding in my frustration. However, we were unable to meet certain deadlines without his cooperation. It affected the performance of the entire team. I had to find a better way to deal with him.

Fortunately, Apple had a mentor assigned to every manager. I could ask my mentor, Fred Foryth, senior vice president for manufacturing, for help, so I went to him.

Fred said, "Kai-Fu, as a manager, you are too weak. Being a manager takes not only wisdom but also determination. I'm telling you now to fire this person within a month."

After talking to Fred, I became a tougher manager. When the older engineer objected to my point at a meeting, I told him 90% of the team agreed with me and there was no need of further discussion. When he was absent-minded at work, I gave him serious warnings. He soon realized I was no longer a pushover. A little over a month later, he resigned before I gathered enough evidence to put him on a performance plan.

When I reported the outcome to my mentor, he said, "I knew it would be hard to fire someone in a month, but I pushed you with an 'impossible' goal to awaken you to your leadership weakness. I saw you as a compassionate leader well liked by most people, but sometimes it takes more than being nice to win people's respect. As a leader, you must effectively execute what's good for the company, and be a good judge on when to put aside your compassion and become a tough manager."

I greatly appreciated his guidance, and was glad about passing his test. At that moment I didn't know I would soon face another test, an even tougher one.

We needed to lay off one person from the team, and I had two names to choose. One of them was a young man who had just come on board and hadn't had a chance to show his talent. The other really didn't work hard enough for our team. He had a house in Nevada, and often took off to take "long weekends." When he claimed to be working from home, he had nothing to show for it. But he was a Carnegie Mellon alumnus who had also been Professor Reddy's student. Professor Reddy had implicitly asked me to take good care of him within my abilities as a manager.

The Carnegie Mellon alumnus heard of the upcoming layoff and figured out he was in danger. He came to my office to beg for my help. He said it might be hard for him to get another job because he was already 40, and that he had two children to support. I felt terrible. However, I knew I couldn't sacrifice the innocent young engineer for his sake. I had to be fair as a responsible manager.

After I laid off the Carnegie Mellon alumnus, he was very angry. He even had a line "laid off by Kai Fu Lee" printed in red on his business card, which he distributed at some conferences later. While I felt uncomfortable, I knew that doing the right and fair thing was more important.

My mentor at Apple understood what I was going through, and he said, "Dealing with layoffs is an important step for a manager to take. You've taken the step. That means you are maturing as a manager."

Layoffs continued to happen at Apple as the company kept losing market share, and those who stayed were unable to receive raises. The morale was therefore low. It was definitely difficult for the managers.

One day, an employee cursed me because his wife and friends had been laid off. He used dirtier words than those in the inner city of New York. I almost exploded. But then I thought of the tough time he was going through, the repressed hurt feelings in other employees and my responsibility to represent the interest of the company. I had to do my job right.

I calmly told the employee, "This is a very hard time for you, for me and for our company. I understand how you feel. If you have suggestions, please tell me after you calm down. We can have a conversation."

A few days later, the employee apologized to me for his rude behavior and thanked me for letting it go. A few years later, he moved his whole family to Europe. He and his wife both found decent jobs. He sends me a Christmas card every holiday season.

Finding Hope in Multimedia
From 1992 to 1996, Apple went through several rounds of layoffs. As financial statements looked worse and worse every quarter, the company had no choice but to trim its work force to save costs.

I was sad to see those laid-off employees silently packing, some of them weeping. It happened again and again.

Even CEO Sculley had to leave. In 1993, Sculley told the board that IBM had asked him to interview for the company's CEO position. All the board members were upset with him because Apple was considering selling itself and IBM was one of the potential buyers.

They asked Sculley to back out from the interview for the conflict of interest. Sculley agreed. But the board ousted him after Apple's deal with IBM failed.

Sculley made a comment upon leaving, "Apple is a lot like Italy. It's a highly creative company, but with that comes chaos."

Sculley was replaced by Michael Spindler, formerly president of Apple Europe. Spindler was a tall and large-framed middle-aged man of German descent, speaking with an accent like Arnold Schwarzenegger's. He looked like a tough man. Everyone thought he was the right person for the CEO position because he had management experience with Apple Europe and around-the-clock work habits that gave him his nickname, "the diesel."

To everyone's surprise, Spindler lost his confidence once becoming CEO. He would sweat every time he gave a speech on stage. Everyone jokingly told one another not to sit in the front rows in order to avoid his sweat and spit.

Spindler had a heart condition and wore his pacemaker to work. He seemed to have a very low threshold for stress. In public, he upheld his "diesel" image, but privately, when he returned to his office, he would cover his head with both hands and hide under his desk when he faced trouble.

It was during the Spindler era that we applied our speech recognition system to a consumer product called Quadra AV. But the product was too high-end. It cost $10,000, which kept it from being popularized. This made me see the ultimate problem with doing software at a hardware company.

The software division of Apple had launched the popular Macintosh OS, which had been superior to Microsoft Windows. But Microsoft as a company entirely devoted to software eventually caught up with Apple's software division. Microsoft's cooperation with the hardware company Intel also made it harder for Apple to compete.

In 1993, Apple launched a palm computer called Newton. Unfortunately, the product received harsh criticism. People laughed at its handwriting recognition. They made a joke about it:

Question: Do you know how many engineers made Newton?
Answer: Fine hungry ("Five hundred" misspelled).

Apple invested more than 500 million dollars in Newton but gained little back. That led to more lay-offs. Just when everything looked discouraging, I found Apple's multimedia technology excellent. I envisioned the possibility of connecting multimedia with the Internet. If we could make some breakthroughs in user interface, I believed multimedia would have a bright future.

With a passion for saving Apple, I wrote a report titled, "How to Revive Apple through Interactive Multimedia" and submitted it to Apple's top management.

Several vice presidents read the report and met to discuss it. Then they decided to accept my suggestion to set up a multimedia department, with me as the director.

Years later, I ran into one of those vice presidents. He said, "We were amazed to read your report. We had only seen you as a speech recognition expert, and didn't expect you to know so much about business strategies. If not for that report, Apple could've missed many later opportunities in multimedia."

After I became director of Multimedia at Apple, my team developed numerous new projects. The first one was to improve Newton. We drastically raised Newton's handwriting recognition rate. But we were disappointed to realize that it was still difficult to sell Newton because of all the bad press associated with the brand name.

Despite our disappointment with Newton, we were happy about succeeding in everything else we did. Our most distinguished achievement came from the graphics team thanks to a creative scientist, Eric Chen. Eric invented QuickTime VR, which could give people a virtual tour of a place. I saw its applicability to museum exhibitions and real estate commercials, so I strongly supported it.

Later QuickTime VR was combined with the movie "Star Trek" to be a CD-ROM video game, which sold a million copies within a month. Today this technology is still widely used in software products such as Google Earth and Google Street View, and often found in the 360-degree views at real estate websites.

## The Road Not Taken

In 1995, vice president of Apple's ATG resigned. Before the company named the new vice president, Dave Nagel, the one who had recruited me, asked me for my opinion on the future development of the ATG.

"ATG is a large team with no specific evaluation guidelines, so it may get a little too relaxed," I said. "If we transform it from a research division into a product department, it can inspire talented team members to brainstorm to make new products. Then ATG's technology may help Apple pull through the current financial crisis."

Dave remained silent. He didn't respond to my suggestion.

A few days later, Donald Norman, a famous psychologist, was promoted to be ATG's vice president. Norman rejected my suggestion. He said it was Apple's tradition to keep research departments and product departments apart. He wouldn't break the tradition.

As our big boss, Dave decided to go for the middle ground between Norman's idea and my suggestion. The ATG would continue to be a research division but would let me take my team to another division, to develop products under another vice president.

However, Norman refused to let me take away my entire team. He said, "Kai-Fu, you should give employees space to make their own choice. Those who want to do products can follow you, but those who want to do research should stay here with me."

He sounded reasonable, but in fact he was making it difficult for me because doing research at Apple was certainly an easier life than doing products. Who would choose to follow me, to face cruel challenges of the market?

I heard Norman had talked to many of my team members and warned them about the risks of developing new products. This worried me even more. What if no one chose to follow me? How embarrassing that would be!

Norman put me in an uphill battle, but I was still determined to win. On a sunny afternoon, I took my entire team to an offsite and gave them a presentation on my plans of product development. I described how combining multimedia and the Internet would create new applications and form a huge space for development. I divided the team members into small groups to discuss how feasible my ideas were and how they could develop their potential better under these circumstances.

I also had the team members play a game. I asked them, "If you were an animal, how would you save Apple?" The team members put on a variety of fun performances. Everyone laughed. It was rare for Apple employees to laugh and hear hearty laughter from one another during that depressing time, so it felt particularly nice.

At the end of the lunch meeting, I said, "I'm not making you choose today. You should follow your heart to make this choice. After all, people have different aptitudes. Some are cut out for research and others for product development. But today Apple is in a crisis. If you can contribute to our cause, I urge you to come together on this journey."

To conclude my speech, I read one of my favorite poems, "The Road Not Taken" by Robert Frost, to the team members:

*Two roads diverged in a yellow wood,*
*And sorry I could not travel both*
*And be one traveler, long I stood*
*And looked down one as far as I could*
*To where it bent in the undergrowth;*
*Then took the other, as just as fair,*
*And having perhaps the better claim*
*Because it was grassy and wanted wear;*
*Though as for that , the passing there*
*Had worn them really about the same,*
*And both that morning equally lay*
*In leaves no step had trodden black.*
*Oh, I kept the first for another day!*
*Yet knowing how way leads on to way,*
*I doubted if I should ever come back,*
*I shall be telling this with a sigh*
*Somewhere ages and ages hence:*
*Two roads diverged in a wood, and I –*
*I took the one less traveled by,*
*And that has made all the difference.*

After the lunch meeting, 90% of the team members chose "the one less traveled by," leaving the research group for the new Interactive Multimedia Department.

Managing the new department was one of my best leadership experiences. Our team created numerous famous multimedia products such as QuickTime and QuickTime VR. Today's Apple multimedia, speech recognition, and even iTune have some roots in what we did. A year later, when Steve Jobs returned to Apple, the team instantly became one of his favorites.

As for the ATG, Jobs simply said, "Our company cannot afford a research institute." Then the entire ATG was closed and all its members laid off.

# The Youngest Vice President

After working in the product division of Apple for six months, I was promoted to be vice president of Interactive Multimedia in fall 1995. I became Apple's youngest vice president at age 33.

I knew quite a number of people had interviewed for the vice president's position, but Apple's financial problems had scared all of them away. In the meantime, Apple employees' resignations reached an all-time high. The COO, CFO, and CMO all left. Out of the 47 vice presidents, 29 resigned. That gave young employees chances of promotion.

As vice president, I led my team to launch the technology of QuickTime on the multiple platforms of the Internet. We produced Apple Media Authoring Tool, developed QuickDraw 3D, and cooperated with the Japanese company Bandai to work on a multimedia product called Pippin, which was similar to the Sony Playstation, but too early for its time.

I also set a series of strategies to make QuickTime adaptable to the Netscape browser, Sun's Java, and SGI's 3D in order for other software developers to cut their use of Microsoft Windows and increase their interest in Apple products. These strategies helped us reach certain goals but didn't meet all of our objectives, given the adverse market conditions.

In the holiday season of 1995, Apple massively produced a lower-end Mac in an attempt to increase sales by lowering prices. However, the plan backfired. Apple ended up with unsold machines at a total cost of two billion dollars. In January 1996, Apple laid off 1,500 people.

In the same year, CEO Spindler lost all of his hopes in Apple. He submitted a resignation letter, in which he said, "In fading away from the place which I loved and feared, I will become whole again, and hopefully renew the father, husband and self I am."

# Leaving Apple

The board of Apple couldn't wait to fill the CEO position. Board members quickly decided to make one of them, Gil Amelio, the new CEO, and agreed to pay him an outrageously high salary for his record of saving National Semiconductor. Amelio took office with an attitude of an expert who specialized in saving companies from collapsing.

On his first day as Apple CEO, Amelio asked me to give him the last 15 minutes of my all-hands meeting, because my division was working on the hottest technology of the company.

Amelio prepared a speech for the hundreds of employees in my division. He said, "Don't worry! This company is doing much better than those I've saved from deep trouble. Give me 100 days. I'll tell you where the company's future is." Then the team had a lot of questions for him, including some pretty tough ones. He handled them quite well.

When I accompanied him walking out of the conference room, I asked him how he felt about my team. He said, "Apple really doesn't have any discipline, no discipline at all."

I didn't expect to hear something like that. Disappointment and anger filled my heart. I found him arrogant. Soon others also saw his arrogance. He asked everyone to call him Dr. Amelio, going against the first-name-basis culture of Apple and other high tech companies in Silicon Valley.

Amelio only worked with his core team on strategies and didn't ask most employees for any input. The strategies he launched 100 days later therefore didn't receive much support. Apple's sales continued to dwindle.

While Apple needed to save costs, Amelio had a luxurious CEO suite built for himself. Many employees were upset with his wasting company budget.

Six months later, Amelio held a company meeting. In a discussion of Apple's failures, he pointed at the audience and said, "Damn it! Don't put me on the spot again!"

He was also disappointing in front of the media. Once he told a reporter, "Apple is like a ship with a hole, and all the sailors are rowing towards different directions. My job is to make these sailors row in the same direction." The reporter asked, "What about the hole?" He was unable to answer.

Under the reign of Amelio, my days at Apple became suffocating. In the meantime many job opportunities arose elsewhere. I learned about them through numerous high tech conferences, where I met many distinguished scientists and entrepreneurs, including then-Sun-CTO (later Google CEO) Eric Schmidt, Netscape founder Marc Andressssen, and leaders from Silicon Graphics (SGI), CEO Ed McCraken and president Tom Jermoluk.

One day, I received a phone call from the SGI Human Resources. The caller said, "We are reorganizing and expanding our company. Our projects include interactive TV, 3D animation and Internet servers. Why don't you come here to take a look? You can tell us what you want to do, and we'll create a division for you."

I felt flattered. Usually companies recruit people to fill existing positions, but SGI was planning to create a position for me based on my expertise and interest!

After a few talks with SGI, I expressed my interest in the Internet. Then SGI promised to reorganize its Internet business and build a new Web Products Division, making me the division's vice president and general manager.

In June 1996, I submitted my resignation letter to Amelio. He said, "You are one of the two best leaders in products, so I'd like you to stay. Just tell me what you want."

However, I had lost all my confidence in Apple by then. I thanked Amelio and insisted on leaving.

In retrospect, I think the six years I spent at Apple gave me lots of opportunities to learn and grow. I will never regret leaving the paradise of academia for Apple, though it put me through merciless challenges of the corporate world like what Adam faces in *Paradise Lost.*

Apple was where I first realized the importance of keeping the user's best interest in mind for product development. Apple has always been a perfectionist in terms of products. For instance, the company was willing to spend $5 more per computer to make a software-driven eject disk function for the convenience of the user. It's Apple's pursuit of perfection that has won so many fans, though it was also that desire for perfection that caused products to become expensive, which once put the company in a severe crisis.

I witnessed how Apple fans remained loyal to Apple products during the company's worst years, and that inspired me to always take consumer interest into consideration in my later career.

Apple also taught me the significance of leadership. A good company cannot thrive without a great leader. After Steve Jobs left, Apple lost its ultimate visionary, cheerleader, and demanding boss. Divisions and departments went on their own, no longer collaborating with one another. Product deadlines were thus often postponed. It was a chaotic situation only Jobs could change.

In December 1996, Jobs returned to Apple. People screamed, "The soul of Apple is back!" Jobs launched drastic reforms, which led to brilliant new products such as iMac, iPod, iTunes, iPhone and so on. Apple once again became a marvelous company beyond the world's imagination.

I missed the miracle by leaving a few months before it all started to happen. But I was happy for Apple, which to me had once been as inspiring as Isaac Newton's apple tree. Apple will always have a place in my heart.

# CHAPTER 7

# The Rise and Fall of Silicon Graphics

On my first day at Silicon Graphics(SGI) in July 1996, there was a fancy little box on my desk. I opened the box and saw a Tag Heuer watch probably worth $1,500. At first I thought the expensive gift was the company's welcoming gesture to an incoming executive. But I soon found out it was an annual present to every employee. Later I also received a lamb-skin jacket. So did every other SGI employee.

SGI was able to afford luxurious presents for employees because its special technology in high powered servers and visual workstation monopolized the markets of graphic design and cinematic special effects. As it contributed to blockbusters such as "Jurassic Park" and "Titanic," its profit margin was naturally high.

Founded in 1982 by James Clark, SGI kept rising and became prominent in the early to mid 1990s. It was once as glorious as today's Google, and actually located in the complex where Google's headquarters currently is.

SGI also had a liberal research environment like Google's. Management would only provide directions and wouldn't interfere with the research process. Engineers had a lot of space to develop their talent.

However, the brilliant SGI engineers could pose a challenge to a new supervisor. Another vice president who had also jumped ship from Apple complained to me, "They gave me a hard time on my first day. Leaving Apple for SGI was like leaving purgatory for hell!"

His complaint made me worry about my first staff meeting. But fortunately, I forgot all the concerns about managing the engineers when I started talking about SGI's future.

"SGI has the best engineers like you to create a bright future," I said. "But there are some future options we have to consider. First of all, should we work with Windows? To use or not to use Windows is a 'to be or not to be' question. Nobody in Silicon Valley likes Microsoft. But not adapting our software to Windows will make us irrelevant. We need to combine our hardware with Microsoft's operating system to ensure our long-term prosperity."

I thought the engineers would argue with me at the end of my speech. But no one did. Instead, one of them said, "Kai-Fu, you are making a lot of sense. We've always thought we should go for the Windows operating system, but management disagreed. They said that would have a negative impact on the sales of our hardware. They wouldn't listen to us."

I learned from the staff meeting that in our area, SGI wouldn't take engineers' input into consideration in terms of business strategies. However, the engineers actually saw the big picture more clearly than management did. SGI products were extremely expensive, unaffordable to the general public. Only specialists in certain fields would purchase them, and that made the company's market quite small. Although it was a niche market, I was afraid it might not last.

I thought of Apple as a precedent. Apple kept its system entirely unique while Microsoft and Intel adapted their products to a shared platform. Of course more and more users turned to Microsoft and Intel products for convenience. In the meantime, a Microsoft-Intel PC only cost $1,000 to $2,000, compared with an Apple computer's $3,000 to $10,000. How could Apple not lose its market share?

SGI also held high prices, and its products were not compatible with the Windows platform, either. As the Windows platform gained increasing popularity, SGI had to take caution not to become another Apple.

I saw two options for SGI: 1) using the Windows operating system; 2) developing a Linux platform to get help from Free Software (a global nonprofit organization advocating open source) engineers worldwide to fight against the Windows-Intel model.

I expressed my view to SGI president Tom Jermoluk, an engineering genius who had worked for Bell Labs and obtained the SGI president's position through his excellent command of technology.

Tom refused to listen to me. He said, "Kai-Fu, don't tell me how good Windows is or how good Linux is. I know the two operating systems better than you do, believe it or not. Would you like to have a technical debate with me?"

I didn't debate with him, because I knew he would be too proud of his own technical knowledge to listen to anything I had to say. I also understood why he rejected the idea of using Windows. Microsoft was deemed a public enemy in Silicon Valley. Anyone cooperating with Microsoft would be considered a traitor.

However, a few years later, after I left SGI, the company did try using Windows and Linux to solve its financial crisis. But it was already too late. SGI eventually had to file bankruptcy. Although it has survived, the glory of the yesteryear was long gone. Looking back, sometimes I wonder if I should have debated with Tom that day. Had I persuaded him to accept my suggestion, would SGI have a different destiny?

## Interactive TV

When I started out at SGI, my division worked on an interactive TV project. Interactive TV allows the viewer to interact with the television set in ways other than simply controlling the channel and the volume. Typical interactive TV uses include selecting a video film to view from a central bank of films, playing games, voting or providing other immediate feedback through the television connection, banking from home, and shopping from home.

Interactive TV requires adding a special set-top box to the existing television set. Other installation and infrastructure arrangements will be added according to the particular approach. Most services offer special programming, news, and home shopping. Some also offer video-on-demand and home banking.

Interactive TV is not as popular today as predicted by many high tech researchers in the 1990s. Back then it was once a hot topic on which numerous high tech companies were conducting research. SGI and Microsoft were among the companies showing an interest in it. Bill Gates mentioned it in his first book, *The Road Ahead,* published in 1996. By then SGI had already developed an Interactive TV business in cooperation with Time Warner.

In the early 1990s, Time Warner persuaded the city of Orlando, Florida to accept a free trial of interactive TV. Then a set-top box was installed in every Orlando family's living room. However, Orlando residents didn't enjoy the new service very much because there were few programs for them to select. In the meantime the set-top boxes cost SGI a great deal. SGI was already tired of the Orlando experiment when I came on board, so the company's top management asked me to figure out a way to cancel it without offending Time Warner or the city of Orlando. I soon found out Time Warner was actually more than willing to call off the Orlando project, but we had to do so delicately, without letting down the subscribers.

I was also assigned to have my division extract some technology from the set-top box to create a new product, hopefully a profit-making one. We looked very hard for treasure in the discontinued set-top box and worked long hours on developing new technology from there. Finally, we created a media server that brought in tens of millions of dollars.

## A 3D Project ahead of Its Time

In 1996, the use of browsers started to become popularized on the American Internet. While Microsoft Internet Explorer and Netscape were fiercely competing for the browser market, I saw sales potential in web servers. The SGI engineers of my division were able to increase the speed of servers, so we launched a series of related products. That brought us two hundred million dollars within two years.

In the meantime, my division was working on a 3D project, which was very similar to the 2009 blockbuster "Avatar." We were dreaming of making the Internet three-dimensional, imagining how lively that would look. We were also developing virtual reality, in which each user would have an on-line ego like an Avatar identity to meet others, play games, drink coffee, and go dancing as if socializing in the real world.

We called our product Cosmo. To launch our 3D web pages, we needed either Internet Explorer or Netscape to bind to Cosmo.

To seek cooperation with Microsoft, I flew to Seattle. I offered letting Internet Explorer bind to Cosmo at no cost. However, the Microsoft general manager I talked to was not persuaded. He said, "Perhaps your technology is better. But we only use Microsoft technology. Even if it's some kind of technology we don't have yet, we'll just let our users do without it."

I returned to SGI, feeling frustrated. But soon good news came. Netscape agreed to bind its browser to Cosmo. At that time, Netscape had 60% market share. That would give Cosmo a lot of exposure. We thought our beautiful baby Cosmo would attract many people right away.

In the meantime we knew the "3D world" vision was still a bit far fetched, and the authoring costs were very high. So we thought we'd make the product for two purposes. We designed two sets of tools that could apply Cosmo to websites. One would make 3D immersive worlds (like Second Life). The other would add animation pictures to two-dimensional websites (like Flash). We advertised the tools with a slogan, "Make the world inside the computer as realistic as the one outside it!"

Those who experienced a demonstration of Cosmo were all amazed to see their virtual egos drinking coffee, learning ballroom dancing, and traveling all over the world.

"Incredible!" one of them said with a sigh. "You guys created a miracle!"

"Microsoft can't do this. Neither can Apple. Nor can IBM," commented another.

However, the highly acclaimed Cosmo didn't sell. We were shocked to find out Internet users were not as interested in animation features on line or three-dimensional websites as we had thought. We had drunk our own Kool-Aid, and let the technical praises get to our heads. I had applied my scientist's head to business – in business, we cannot measure success by what is cool and new, but by what users vote with their wallets. We also learned that by trying to cater to both types of users, we didn't do a good enough job to support either.

Given the sensational popularity of "Avatar" in the holiday season of 2009, I wonder if launching Cosmo today would be a completely different story. Unfortunately, we made Cosmo in the late 1990s, when the general public just began to use the Internet. The market was not mature enough to generate a need of virtual reality.

The Saddest Spring of My Life
In spring 1998, SGI was no longer the prosperous company it had been. Its sales dropped and financial resources tightened. So, as we started our planning for the second version of Cosmo, to recover from our mistakes, we realized that we might not get our second chance.

A new CEO, Rick Belluzzo, came from HP, where he had established a sales record of HP printers by selling the machines for below-market prices and making a profit from the ink. As he used to be an accountant, he was good at numbers and made decisions based on numbers.

When Belluzzo saw the cost of my division exceeding the income we generated, he told me that he was going to eliminate the entire Cosmo division.

I was shocked to hear the horrible news. How much this would hurt the more than 100 hard-working employees in my division! I couldn't just sit there and let it happen, so I said, "If you are unwilling to continue with this division, I can sell it to another company."

"OK," Belluzzo responded with an accountant's attitude. "How much do you think that can make?"

"About $15 million," I gave an optimistic estimate.

"How much have you spent in the past two years?" he asked.

"Around $20 million," I replied.

"That's all right. We would still take a loss, but it's better than getting nothing out of it," he said. "Sell it then. I'll give you four months to sell not only the division but also all the engineers."

I had no choice but to start looking for buyers. I flew to New York, Chicago and Japan. Sony expressed a strong interest in the division for the company's exploration in 3D animation. They agreed to pay $15 million, and that made me feel relieved, thinking I had found a good new employer for the engineers in my division.

However, just before we signed the contract, Sony backed out.

At this point I only had one month left to sell the division. What was I supposed to do? I had once thought of converting the division into an independent new company, but that required a lot of funding, which I was unable to get.

Finding another buyer seemed to be the only way. Pressured by time constrains, I went to Platinum Software, which had once given us a low-ball offer. Platinum Software knew I was there because our deal with Sony had failed. The company took advantage of our desperate situation by offering only five million dollars. That put me through a painful struggle. But in order to save the engineers in my division from losing their jobs, I signed the contract.

With a heavy heart, I held a staff meeting to announce the buyout. I tried very hard to suppress the pains in my heart and said, "I believe you all have been aware of our less than satisfactory sales. Per our new CEO, I began to try selling our division to another company four months ago. Now Platinum Software has decided to buy the division."

I saw a small stir in the audience. But it soon calmed down. Apparently the engineers knew having a buyer was better than not. This way at least they wouldn't be laid off right away.

No one questioned the buyout. All the engineers quietly accepted the fact. Their silence somehow felt suffocating to me.

## Tormented by Guilt

I blamed myself. I thought should have found a better buyer for the SGI division, especially after I heard what happened later. Unfortunately, Platinum Software sold itself to Computer Associates after buying the SGI division. Computer Associates was not interested in 3D technology. After the merger, the company only kept 10 of the division's engineers. More than 90 of them were eventually laid off just as they would have been from SGI.

The news shocked me. That meant I indirectly made more than 90 engineers lose their jobs! I knew many of them were breadwinners of their families and had children to support. How were all those families going to cope with the sudden loss of income? It hurt me to think about what they were going through, but I couldn't stop thinking about it. I became another person, broody and quiet. Sometimes I sobbed uncontrollably.

My abnormal behavior worried my wife. She said, "I think you may have depression. You should see a doctor. You can't go on like this!"

When I went back to Taiwan on vacation, my fifth sister, a social worker, recommended me to see someone who could help me. During my first visit, the doctor said, "You need an emotional outlet. So I'll pretend to be one of those engineers, and you tell me whatever you want to say to them."

I closed my eyes, imagining it was the engineers in front of me. I tried to talk but was only able to say, "I'm sorry--- " Then I burst crying.

The doctor just let me cry. Later, when I calmed down a little, I started to tell him all about my frustration and sense of guilt.

The catharsis gradually relieved me of self torture. While still feeling guilty towards those engineers, I was determined to learn from the experience and become a better leader to my future subordinates.

My two years at SGI were filled with learning:

I learned that it is not innovation alone that matters, but innovation that is useful.  The painful experience with Cosmo taught me to humbly understand user needs, before letting the scientist in me get carried away with cool new technologies.
I learned that the development of a new product must be focused.  It is not right to try to be everything to everybody. Our attempt to cater to "website animation" and "immersive 3D" at the same time was a strategic error.
I learned how to manage a business and a P&L (profit and loss) center, how to motivate a sales team, and how to develop a new business.
I learned the difficulty of leadership. Having to seek an acquirer at one meeting and motivate the troops at the next can be disconcerting. But to lead a team, it is important to multitask, and to switch contexts on a dime. Through the trials and tribulation of the final phase of Cosmo, I became prepared for the most difficult challenges for a manager.

# Going to Microsoft

While I was still working on selling my division of SGI, I received a call from Intel. The caller said, "We are considering building a research center in China. Would you be interested?"

That reminded me of my teaching experience in China, and of my desire to help Chinese students, so I asked, "How big would the research center be?"

"Probably a few dozen people," the caller replied.

"Um, if it's a bigger team, perhaps I'll be interested," I said.

"Really? That's great!" The caller sounded pleasantly surprised. "I'll tell my boss about it. I didn't expect you to be interested.

China in 1998 was certainly far less appealing than it is today. It was considered backward then. But to me, China's underdeveloped research environment only meant it was in need of my help. I was more than willing to go to China, especially knowing that would make my father happy. But I was concerned about Intel being a hardware company. I was getting tired of working for a hardware company as a software specialist, because the hardware companies I had worked for didn't really understand the needs of software engineers.

When I hesitated, Intel offered me a high position as CTO of Intel Asia, and managing director of the China research center. The company kept pushing me to make a decision, implying there were other candidates. However, I needed more time because I was still working on selling Cosmo.

Before I finally sold the SGI division, I saw Microsoft as a potential buyer and flew to Seattle. The Microsoft vice president I spoke with turned me down. But I ended up with an unexpected gain from the trip.

At a restaurant near the Microsoft headquarters, I had dinner with Xuedong Huang, who had been a post doc under my supervision at Carnegie Mellon University and was then working for Microsoft. I told him that I would leave SGI after selling my division and might take an Intel offer to work in China.

Xuedong was surprised to hear my plans. He said, "Since you are willing to go to China for Intel, what if Bill [Gates] has a similar idea? Would you work for Microsoft in China?"

"Do you have a timetable for that?" I asked. "And how big would the China team be?"

"Our current project may be too small for you," said Xuedong. "But I think you should have a talk with Nathan (then-Microsoft-CTO Nathan Myhrvold)."

"Oh?" I expressed my uncertain feelings. "I've had two business negotiations with Microsoft, which showed me a strong-headed, inflexible image of the company. I'm not sure if I would fit in."

"What outsiders see is not the real Microsoft," Xuedong said. "Besides, your job would be in a research center, and your boss would be Rick Rashid, your former professor at Carnegie Mellon. Rick is modeling Microsoft Research after Carnegie Mellon. I bet you will like the culture of the research center."

I thought exploring the option wouldn't hurt, so I said, "OK, please tell Rick and Nathan I'd be happy to talk to them."

After returning to Silicon Valley, I continued to look for buyers for my SGI division. I almost forgot my conversation with Xuedong until receiving a phone call from Dr. Rick Rashid, the one who had recommended me to the Ph. D. program in Computer Science of Carnegie Mellon University.

"Microsoft Research is based on the Carnegie Mellon model. It would be the right place for you," said Rick, a second-generation Lebanese American with a very gentlemanly demeanor. "If you join us, we will make the China center bigger and better."

"How much bigger and how much better?" I asked directly.

"How much bigger and how much better would you like it to be?" Rick asked with excitement in his voice.

"Can it be as big as the Cambridge center?" I posed a specific question. "Microsoft announced to have invested $80 million in the Cambridge center for six years. Can Microsoft Research China receive so much investment?"

"Of course!" he immediately gave me a positive response. "The investment will be the same. You can hire 100 people first and we'll see how it goes. If it works well, we can add more people."

I was surprised to hear no hesitation from him. This really sounded fantastic!

I took the next step to see Nathan, who reassured me that the research environment would be like Carnegie Mellon's.

"Kai-Fu, the new technologies you developed at Apple and SGI totally amazed us," said Nathan. "But haven't you noticed that those companies never had enough patience for you to develop products? They gave up too soon on the products the market wasn't ready for, whereas Bill [Gates] and I have more patience. We can wait for new technologies being incubated. Let's see, would you like to let your creative inventions continue to be put aside? Or would you like to change the world?"

He seemed to know what spoke to me most persuasively, and apparently, he could tell I was being convinced. He suddenly changed the subject, "By the way, Kai-Fu, I have a super computer at home. Do you know how I cool it off?"

Nathan is a not only exceptionally smart but also very interesting person. Later he left Microsoft to travel around the world, learning the art of cooking everywhere he went. He also founded Intellectual Ventures, a company buying patents and making a profit from them. Sometimes it helps Microsoft negotiate patent deals to obtain lower prices than Microsoft could get on its own. By now Intellectual Ventures has become the company with the largest number of patents.

That day in Nathan's office, his casual tone created an easy atmosphere for me to imagine developing the most cutting-edge technology in China and helping striving Chinese students grow with it. Wouldn't that be a dream come true to me?

I also thought of what had happened to my Casper and Cosmo. That made me long for joining a software company with a real understanding of software development.

Microsoft could be the software company meeting my needs, I thought. And the company made me feel special by telling me that Bill Gates would personally make an effort to recruit me.

Gates became the richest man in the world in 1995. By the time I was about to see him in 1998, I had read a lot of media reports about his wealth. I was inevitably curious about what I was going to see at my meeting with him.

When I stepped into his office, it didn't feel like entering the richest man's territory at all. There was nothing luxurious. The oak furniture looked tasteful but understated.

Gates was in his early 40s then. Despite being in charge of the world's largest software empire, he was wearing messy hair and a polo shirt like a typical engineer.

He asked me about my hopes and concerns about Microsoft Research China with a gentle voice and friendly attitude. I sensed his strong interest in the China market.

I accepted the job offer. It was a level-15 position, only a step lower than a vice president. My title was managing director of Microsoft Research China.

Leaving Silicon Valley
After signing my contract with Microsoft, I returned to my house in Silicon Valley. When my wife saw me walking in, she said, "Someone named Steve from Apple called you. He asked me to make sure you'll call him back."

"Steve?" I was stunned for a minute. "I don't know a Steve from Apple."

"Really? I thought he was your friend," my wife seemed a little puzzled, too. "He asked me what you were up to, and I told him we were going to China. We talked for about 15 minutes."

"Oh?" I felt even more confused. "Steve who?"

"He didn't say," my wife said.

I called the number Steve had left with my wife, and the person who answered was Steve Jobs! The co-founder and savior of Apple!

What made Jobs dial my home phone number after his successful comeback? I couldn't wait to find out. When we were on the phone, he said, "Why don't you come back to Apple?"

"I never thought about this, Steve," I replied. "It's been two years since I left."

"Listen, that doesn't matter," said Jobs. "All your former subordinates said you were a good boss. They asked me to bring you back. Why don't you come back here to take a look before going to Microsoft?"

"That sounds very nice, but I already accepted Microsoft's offer," I said. "I'm sorry!"

Although I didn't go back to Apple, I appreciated the invitation, and was impressed by Jobs. He was an entirely different person from the arrogant tyrant portrayed by the media when he talked to me on the phone. He sounded friendly, understanding, and eager to recruit talents. He made me believe he would take Apple into a bright future.

After resisting the last temptation from Apple, I started preparing to leave Silicon Valley, where my wife and I had established a beautiful home.

When we sold our first house in Pittsburgh and moved to Silicon Valley, we thought we were going to settle down. Given the most advanced computer technology in Silicon Valley, I assumed changing jobs would still keep me in the same area. Thinking this way, we devoted all our energy to remodeling our newly bought house on the hillside of Saratoga, California.

Beside the old house was a small cottage converted from a stable. In order to completely remodel the old house, my wife and I slept in the cottage first. We decided to redo the flooring with marble, and wanted the 1,000 square pieces of marble to look like a natural whole when combined. All the contractors we had talked to said that would be too much to ask. None of them took the job, so we did it on our own. Every day we studied the patterns of the marble pieces and figured out how to make their patterns connect when putting them together. After countless hours, the living room floor really looked like a gigantic slab of marble from Mother Nature. We both lay down on the elegant, cool floor, feeling satisfied but too exhausted to move.

*Holding my daughter Jennifer at the entrance of our Silicon Valley home in its remodeling process*

When the main house was ready for us to move in, everything inside looked brand new and nearly perfect. Then we happily began to grow plants in our yard. We especially liked growing roses in various colors and enjoyed their gorgeous blooms in spring time.

It was hard for me to leave such a sweet home, and perhaps even harder for my daughters. The 7-year-old Jennifer and 3-year-old Cynthia were raised in this serene, spacious environment. Would they be able to deal with China's crowdedness? Would Jennifer be sad to leave her friends? Would Cynthia cry about leaving the neighbors who adored her? I worried about taking them to a different world, but at the same time I was hoping that with my adventurous genes in them, they would embrace new challenges.

*Carrying my youngest daughter Cynthia on my back in our Silicon Valley home*

I felt like sharing these thoughts with my wife, but I was almost afraid of asking how she felt about going to China, where living standards were lower. Another woman probably would have opposed the move or at least complained about it. My wife, however, simply said, "Home is where-ever our family is. The kids and I will follow you where-ever you want to go."

I couldn't thank my wife enough for being so supportive. She removed all my concerns and empowered me to go ahead, to pursue my dream, which was looking more and more real: I would create a world class research center in China!

*My family in our Silicon Valley residence before moving to China*

*Seven-year-old Jennifer (right) and three-year-old Cynthia in Beijing in 1998*

# CHAPTER 8

# A Mission Impossible

As my new boss at Microsoft Research, Dr. Rick Rashid told me a funny little story:

When Rick was recruited from Carnegie Mellon University to found Microsoft Research in 1991, Microsoft Corporation wasn't the software giant it later became. A close friend of Rick's didn't even believe Microsoft could last five more years. The two of them bet 25 cents on Microsoft's future.

Rick won tremendously more than the 25 cents. By the time he talked me into founding Microsoft Research China, Microsoft Corporation had evolved into one of the most influential companies in the world, and Microsoft Research had attracted countless distinguished scientists, including winners of the Turing Award, the Fields Medal and the Wolf Prize.

Microsoft Research has a mission statement titled, "Turning Ideas into Reality." The mission statement describes Microsoft Research as follows:

Microsoft Research is dedicated to conducting both basic and applied research in computer science and software engineering. Its goals are to enhance the user experience on computing devices, reduce the cost of writing and maintaining software, and invent novel computing technologies. Microsoft Research also collaborates openly with colleges and universities worldwide to advance the field of computer science.

Unlike most other companies' R& D, Microsoft Research doesn't set timelines for product development. It lets scientists take all the time they need to do long-term research. But interestingly, it keeps showing higher investment returns than those doing product development under time pressure.

When I joined Microsoft in summer 1998, I heard people call Microsoft Research a "Redmond miracle," given the location of the Microsoft headquarters in Redmond, Washington, a city near Seattle.

My job was to duplicate the miracle in China. I was more than enthusiastic about it, though most people around me said it was a mission impossible.

"It'll just be Microsoft's little toy," said one of them. "There's not much you can do in China."

"Don't go! It doesn't have much of a chance," said another. "Where are you going to find top-quality Ph.Ds in China? You have to guide them like showing a toddler how to walk. Imagine how tiring that would be!"

These friends saw me as a Don Quixote, full of admirable ideas but doomed to fail.

To prove them wrong, and more importantly, to realize my dream, I went ahead to start building my team. I first targeted Microsoft researchers of Chinese descent, who accounted for just under 10% of the research staff in the headquarters.

I sent all of them an e-mail, saying, "I will open a new research center in China. It will be as great as Microsoft Research Redmond and Microsoft Research Cambridge. If you are willing, you can lead Ph.Ds in China onto the path of innovation. This pioneering work will bring us a tremendous sense of achievement!"

I also went to see some of them, trying to recruit them in person.

They all seemed interested when I talked about the prospect of Microsoft Research China. But as soon as I asked them to go there with me, they showed hesitation.

"Kai-Fu, this is a great idea, but my wife and children are used to living in America," said one of them. "I don't think they'll agree to move to China."

"The environment of China is polluted and crowded," commented another.

"Moving is too much trouble," said another. "I'm really not hungry enough to do this. I'm happy where I am."

Every time I faced a rejection, I told myself not to give up. I believed I could build a world class research center if only five senior researchers followed me.

What if I failed to find five such people? In that case I could at least establish the best research center in China. I believed in the intelligence of Chinese students. More importantly, they were hungry for knowledge and willing to work unbelievable hours. I would be able to help them learn and make them excel in a few years. If a few years were not enough time, I would give it 10 years. In 10 years there would be a huge difference. Sooner or later, I would fulfill my father's last wish for me, to help young Chinese people open their horizons.

With such determination, I looked beyond Microsoft Research for team members. Then in Microsoft Corporation I found a software testing manager, George Chen, and a senior software engineer, Xiaoning Ling.

George was about 37 years old. Born and raised in China, he came to America for graduate school and earned a Ph. D. from Washington University. While he enjoyed his new life in the United States, he always kept China in mind.

To George, the Microsoft Research China plan sounded like a great opportunity that would enable him to give back to his homeland. He said, "Kai-Fu, I'll do anything for you in China. I've sold insurance there. I'm assertive by nature. I also know many professors and department chairs in China's universities. I can help you build connections with those universities. I can go anywhere you need me to go. I'm not afraid of working too hard."

His strong motivation moved me and boosted my confidence in team building.

Xiaoning also came from China. He was born in the 1950s. The Cultural Revolution from 1966 to 1976 forced him to leave school. He was assigned to work in a steel factory, but he spent all his spare time studying electronics on his own. He taught himself to put together a semi-conductor radio and learned about computers. After the Cultural Revolution, he passed an entrance exam to enter Peking University, the most prestigious university in China. Later he pursued graduate studies in America and finally became a software engineer at age 40.

I met Xiaoning through a friend. When I asked him, "Would you consider going back to China?" He immediately said, "No need to consider. I've already decided to go back."

Later I learned that his family remained in the United States for a lifestyle preference. He endured separation from them for the sake of Microsoft Research China.

The three of us plus our secretary, Eileen Chen, formed the original team of Microsoft Research China.

*The founding team of Microsoft Research China, from left to right: George Chen, me, Eileen Chen, Xiaoning Ling*

## "Micro$oft" in China

In the late 1990s, Microsoft was often purposefully misspelled as Micro$oft in China because the software empire looked incredibly wealthy to the Chinese living in a developing economy. Many young Chinese regarded Bill Gates as an idol after he gave a speech at Tshinghua University in 1997.

Later in the same year, however, the Department of Justice accused Microsoft of breaking anti-trust law by bundling browser software with the Windows operating system for sale. The headline news spread to China. Somehow it made Microsoft's monopoly look even worse than the rampant pirating in China to the Chinese. While China's hardware sales increased 30% in 1998, the country's software sales dropped 30% as a result of the people's negative feelings against "Micro$oft."

That was when my four-member team started to establish Microsoft Research China.

After renting the sixth floor of a highrise building in Beijing as our site, we went to one university after another to recruit talents. We demonstrated the most cutting-edge technology in front of students. They saw us create a three-dimensional model of a person and then made the model laugh. They saw me pose like an orchestra conductor to a computer and the computer follow my gestures to play beautiful music. They saw me speak into a computer, and words would come out on the screen.

As the science-fiction-like demonstrations astonished the students, I told them that joining Microsoft Research China would make them part of the miracle-creating team. They responded enthusiastically. More and more, I felt the impossible becoming the most likely.

In the meantime, I rented a single family house in a Westernized community where there were a gym and an amusement park. I hoped the environment was similar enough to an American residential neighborhood to make my children feel at home right away.

On Aug. 29, 1998, my wife and daughters arrived in Beijing with luggage weighing more than 1,000 pounds. We moved all our belongings from Silicon Valley for a long stay in China.

I bought one-way tickets for all of us. I had no back-up plans. The only plan was to succeed.

On Nov. 5, 1998, Microsoft Research China announced its establishment at a press conference held in the Beijing International Club. We told the media that we had an $80 million investment for the next six years and would expand to a 100-member team.

We attracted a lot of media attention and more than 300 guests, including computer scientists from China's National Science Institute, university presidents, deans, professors, government officials and American diplomats. Bill Gates sent a video recording of his speech for the opening ceremony.

In my speech at the opening ceremony, I said, "It's lucky of a software engineer to work for Microsoft. It's also lucky of a Chinese American to work in China. And it's definitely, incredibly lucky of a researcher to create a research center, to lead people climbing the peak of wisdom. So today I'm triple lucky! I'll cherish my luck and make the best of it!"

That was four months after my first day with Microsoft. But I still had none of the five senior researchers I needed.

## New Blood

Soon after the establishment of Microsoft Research China, a recruiter contacted me and sent me a resume of Ya-Qin Zhang, who was a director of Sarnoff Laboratories in the United States.

The resume immediately impressed me. It presented a genius who entered China's Science and Technology University at age 12, obtained a Master's degree at age 20 and earned a Ph. D. from George Washington University at age 23 with a dissertation that received the only perfect score in the university's history.

I was overjoyed to hear such a distinguished scientist was willing to join Microsoft Research China. I made an international phone call to him and invited him to Beijing for an interview.

When I first saw Ya-Qin in the community center of my residence, he smiled at me and greeted me like an old friend, "Hi, Kai-Fu!"

We exchanged our opinions on developing advanced technology in China. Ya-Qin quickly said he would join Microsoft Research China no matter what compensation package we could offer. I was touched and felt empowered at the same time, foreseeing a snowball effect after he came on board.

Indeed, Ya-Qin set an example for other outstanding Chinese scientists to follow. We soon recruited Jin Lee from Sharp Labs and Shipeng Lee from Sarnoff Labs. Both of them greatly admired Ya-Qin and happily agreed to join us because Ya-Qin was with us.

I also persuaded a researcher of the Microsoft headquarters and fellow alumnus of Carnegie Mellon University, Harry Shum, to transfer to Microsoft Research China. Besides, we reached a multimedia expert and research director of HP Labs, Hongjiang Zhang.

We looked for talents in China as well. As a result, Professor Jian Wang of Zhejiang University and Professor Changning Huang of Tsinghua University left their teaching positions for us.

We formed the first management team of Microsoft China. These elusive senior researchers were starting to come in! Our next task was to hire researchers. This posed a challenge because computer science Ph.Ds in China at the time were generally far behind their counterparts in the United States. If we used American standards to evaluate them, we might not find anyone qualified.

*My core team at Microsoft Research ChinaFrom left to right front row: Hongjiang Zhang, Kai-Fu Lee, Xiangyang Shen, Ya-Qin Zhang; middle row: Shipeng Lee, Jiang Lee, Jin Lee, Wenyin Liu; back row: Eileen Chen, Jing Chen, Sheila Shang, Jian Wang, Eric Chang.*

To solve the problem, we created a new system. While Microsoft Research never had an assistant researcher's position, I made the Ph.Ds accepted to Microsoft Research China assistant researchers. They were given two years to do research recognized at the international level and get promoted to be researchers. If they failed to meet the expectations, they would have to look for another job.

We also designed a competitive compensation package for each researcher and assistant researcher, though we couldn't make it exactly the same as what the headquarters offered researchers there. I recalled being blown away when visiting the luxurious mansion of a headquarters senior researcher's in Redmond, Washington. When it was time for dinner, he opened his back door and invited me to board his boat. We sailed across the lake behind his house and ate at a gourmet Italian restaurant on the other side of the lake. After that, we sailed on the lake for sightseeing while drinking exquisite red wine and smoking fragrant cigars.

This researcher obviously had made a fortune from Microsoft's stock in the early to mid 1990's. Microsoft in 1998 was no longer able to give out that many shares. But I still managed to give each assistant researcher of Microsoft Research China a higher amount of stock and salary than any of the other international corporations in China would offer.

The excellent compensation package and the prestige of Microsoft attracted countless applicants. To select the best of the best, we gave them a written exam first, and interviewed everyone with a score exceeding our minimal requirement. But to my surprise, 90% of the test takers failed.

As we held written exams in all the cities where there were applicants, our university relations manager, George Chen, flew from city to city. He insisted on giving every applicant a chance, so he would even go to a city where there was only one applicant.

In Wuhan, a riverside city in central China, a young woman named Qian Zhang found herself the only test taker and was asked to go to George's hotel room to take the test. She suddenly felt suspicious.

"Is this a scam? What if the administrator is a rapist?" she thought.

But soon she realized there was no danger. George walked out of the room right after handing her the exam and didn't return until the time to collect it. Qian passed the written exam as well as a few interviews. She became one of our assistant researchers. Today, she is a full professor at the Hong Kong University of Science and Technology.

## KFC, Poker and a Black Guerrilla

Microsoft Research China needed researchers with creativity and problem solving skills, so we asked unconventional interview questions including:

- Why is the manhole cover round?
- Please estimate how many gas stations there are in Beijing.
- What would you do if you had a disagreement with your adviser?
- Give an example where you solved a very difficult problem.
- Here are two unevenly-shaped ropes. It takes one hour to burn each. Use them to measure 45 minutes.

Those who answered such questions well are all quick thinkers. Microsoft Research China gave them all the freedom and time they needed to develop their potential.

To benefit more people, Microsoft Research China took college students as interns in the summer. If the interns achieved research results, they could publish papers in international science publications, and that would help them get accepted to top graduate programs of American universities such as MIT, Stanford and Carnegie Mellon.

Chinese students were raised in a hierarchy-oriented culture and would generally present a "yes sir" attitude towards their teachers. But I decided to Americanize their manners in the research center in order to create an egalitarian environment for everyone to thrive. While the students thought they were supposed to use "Director Lee" or "Dr. Lee" to address me, I insisted that they call me Kai-Fu. Some of them were not used to it at first but gradually adapted to it.

Given the casual atmosphere, everyone felt relaxed and often teased each other for fun. Ya-Qin began to call me KFC based on my initials K.F. To get even, I called him "toothpick," which is *ya-qian* in Chinese and sounds close to Ya-Qin.

When it came to key issues, I always gathered input from the entire staff. I once sent out an email requesting them to brainstorm for naming our six conference rooms. I told them I would name the first one Gunpowder Warehouse because the Chinese invented gunpowder and I hoped to see more sparks of innovation coming out of the conference room. Then I asked them to come up with more names and vote for the best names.

Many names came out. Then everyone voted. In the end the six chosen names included Gunpowder Warehouse. Three of the other five names were also based on Chinese inventions. They were Compasses Hall, Paper Mill and Printing Factory. Another name, Abacus Room, was math related. There was also a Zero Room, coming from the mathematical concept of zero.

The names were going to be printed in both English and Chinese. At first we had trouble translating Zero Room into Chinese because it could be translated as *ling-tang* and would sound like "funeral site" in Chinese.

Fortunately, two of the assistant researchers discussed it and found a brilliant translation: *ling-gan-wu,* which means the house of inspiration.

We all felt closer to one another when the names we had created together were posted on the doors of the conference rooms. That was exactly what I meant to do, to build a team spirit. For the same reason I suggested playing cards together. To create more fun, we had the loser crawl under the table.

Although my strong skills in bridge helped me pick up Chinese card games easily, I lost once in a while and had to crawl under the table. That made the students laugh the hardest because they had never seen a leader, supposedly authoritative in Chinese culture, perform a clown act like that.

Everyone became more and more relaxed with me. On my birthday, they decorated my office with balloons and gave me a stuffed animal, a black guerrilla with a sticker on his tummy that read, "Push."

I pushed it and heard recordings from each one of them. Many of them were making fun of me in a way they would never do to another Chinese leader.

*My birthday present from employees of Microsoft Research China*

# Ubiquitous White Boards

Whenever we had visitors at Microsoft Research China, they said to us, "How come you have so many white boards here? Even the top of the coffee table is a framed white board!"

I created a white board culture based on my Carnegie Mellon experience. In 1983, Professor Reddy and I often drafted our ideas on a white board during our discussions. I found it the best way for engineers to communicate.

The white board culture of Microsoft Research China represented open-mindedness. It treated all the staff members as equals, let them express whatever thoughts they had, and allowed them to make mistakes, which could be easily erased. It also symbolized teamwork, inviting staff members to brainstorm as a team.

Ya-Qin was as much of a white board fan as I was. Before he came on board, he asked me for a big white board in his office.

I said I would give him a 14-meter-wide white board, but later I realized my metric system math was flawed. Ya-Qin ended up with one four meters wide. Then he often jokingly said to me, "You owe me 10 meters!"

To maximize the effect of the white board culture, I designed "white-board coffee tables." Each tabletop had an inlay of white board. Even a round tabletop had an inlaid white board in the same shape. Visitors often said they had never seen a round tabletop white board.

*Ubiquitous white boards at Microsoft Research China*

If our staff members suddenly came up with a good idea during their coffee break, they were able to jot down the idea right away. The white-board coffee tables and the sofas around them also provided a comfortable setting for them to have discussions.

Once an American professor looked at the white-board coffee tables during his visit and said to me, "This is a really great idea! Kai-Fu, would it be possible to send one of these to America to me?"

## "How to Say No to Your Boss"

"As a Chinese American, are you culturally more Chinese or more American?" Many journalists have asked me this question.

Before 1999, I always answered that I knew both cultures well. But an incident in 1999 made me realize I had been overconfident about my understanding of Chinese culture.

On Jan. 31, 1999, Ting Liu took the train from Harbin, a city in northeastern China, to Beijing for his interview at Microsoft Research China. He thought I wanted him to take the train in order to save company budget, though I said the research center would pay for his trip, either airfare or a train ticket.

When someone later told me about Ting's misinterpretation of my words, I almost couldn't believe my ears! Since I grew up in America, I was used to direct expressions of thoughts. I didn't know the Chinese might have different interpretations of a simple sentence.

According to the staff member explaining the misunderstanding to me, when Ting asked me whether he should take a plane or train, he really was asking if we would pay the higher expense of airfare. And had I been willing to pay for the airfare, my answer should have been "plane."  But the answer "either way" sounded like a way to decline his request without saying no! This was just too much psychological acrobatics for me!

I began to wonder if I didn't truly understand what the staff members were thinking because of my different background from theirs. I worried about it more after learning that Chinese employees would never say no to their boss.

One afternoon I held a meeting with the entire staff. First, I asked Xiaoning, who had worked at the Microsoft headquarters for years, to explain American corporate culture. He wrote on the board a topic that surprised everyone in the audience, "How to Say No to Your Boss."

After his speech, I said, "You have the right to say no. You are working for the company, not for your boss. You should believe you know more than your boss in your specialized field. You should think Kai-Fu doesn't know everything."

I told the story of the airfare and train ticket. Then I said, "I hope you all frankly tell me what's on your mind. I'd like you to see the research center as your home and your colleagues as your family."

My sincere request encouraged one of them to speak up. He said, "Kai-Fu, to tell you the truth, as assistant researchers, we always worry about not getting promoted in two years. That makes it hard for us to concentrate on our research."

"I understand this concern you all have," I immediately responded. "But you have to put that aside and work hard if you want to stay with us."

Obviously this answer didn't solve the problem for the assistant researchers. Another one said, "Kai-Fu, you are talking from a supervisor's point of view. Please put yourself in our shoes. We got Ph.Ds from prestigious universities. But now we are temps at Microsoft without knowing where our future is. Please imagine how that would feel!"

"Sure! Thank you! This is the kind of communication I'd like to have with you," I said. "Sorry for my inappropriate previous answer. Let me introduce our system again. It doesn't necessarily take two years for you to get promoted. For example, Yingqing Xu has impressed us with his innovative ideas and team spirit. We were going to promote him this month, but let me take the opportunity to tell you right now. Hopefully he'll be your role model and you will all be promoted soon!"

"We don't have a quota for our researchers," I added. "If everyone does a good job, everyone will get promoted."

Their eyes all sparkled as soon as they heard this announcement. I should have communicated this earlier, but in a traditional Chinese environment, it was not management's responsibility to be transparent, and certainly the employees did not feel they had the prerogative to ask for clarification or transparency.

After this exchange, I felt the communication really working. But soon I saw doubtful expressions on some of their faces. I realized they were still concerned about one thing---what if they couldn't get promoted?

I addressed the concern, "Although I cannot promote you if you don't qualify, I'll give you a certificate proving you've done post doc work with us, and we'll give you enough time to find another job. So at the very least, your experience here won't hurt your career!"

Everyone looked relieved upon hearing this explanation. The atmosphere became casual. Some of the assistant researchers began to feel comfortable enough to make complaints.

One of them said, "Our manager gives us too much pressure. We can work day and night for a short period of time, but we can't take it long-term. It's too much stress!"

Another said, "Some managers always tell us how hard they used to work and how hard Kai-Fu used to work. But working 16 hours a day is really too much! Our brains are numb after that!"

"You are right! We can't expect unreasonable overtime." I acknowledged their frustration. "I'll communicate with all the managers and make sure they respect your private life. Besides, I hope you are all doing something you love and are good at, so you will naturally work hard and achieve great results!"

The meeting helped to shape an open culture in the research center. All the staff members began to speak their minds. That brought them closer together and enhanced their teamwork.

To have regular communication with assistant researchers, I scheduled a lunch with all of them every other week.

## Useful Innovation

What Microsoft Research generally does is basic research, which means working on theories that are too new for immediate applications. Former Microsoft CTO Nathan Myhrvold once said, "The theoretical research we do may see results in 10 years. Some of it won't see results until 100 years later."

It's admirable of Microsoft to invest in the future of mankind without expecting short-term financial returns. It's wonderful that the company can afford to do so. However, in my heart I knew that Microsoft Research China must not take the same route as its headquarters in America or the Cambridge center. Compared with the many distinguished scientists in those two centers, we only had six established researchers, including me, and 20 fresh graduates from China's Ph. D. programs. It was unlikely for us to publish as high-quality papers as theirs.

How were we going to be compatible with them then? I kept thinking about this and repeatedly discussed it with my management team. Our conclusion was to create a unique branch of Microsoft Research by doing research that would lead to useful applications sooner. Instead of looking at 10 years, we aimed at applying our research results to consumer products in three to five years. We would initiate open dialogues with product groups, try to understand their business objectives, and use those to guide our work.

When we started a project, we had to know what it could be applied to and how it could benefit users if it succeeded. When we evaluated a project, we looked beyond how innovative it was to see what value it could bring to users. I would not repeat my mistakes at SGI – it is not innovations that matters, but useful innovations.

With clear objectives in mind, we formed six research teams:

- Internet Multimedia--- led by Ya-Qin to study the transmission, compression, and coding of multimedia content.

- Multimedia Computing---led by Hongjiang to do multimedia categorization, such as face recognition, video segmentation, and multimedia classification.

- Image Computing---led by Harry to work on 3D virtual reality in order to create interactive multimedia.

- Multimodal User Interface---led by Jian to design new user interface, such as new ways of entering Chinese phonetic symbols and new digital pens.

- Natural Language---led by Changning to enable computers to teach the user English and correct the user's grammar while building a large database and a statistic model.

- Chinese Speech Recognition---led by me to adapt the technologies of speech recognition and speech synthesis to the Chinese language.

Since Microsoft Research China was understaffed, I decided to lead a project while working as managing director. I didn't mind being overloaded for helping the research center grow.

There were tremendous challenges in adapting the technologies of speech recognition and speech synthesis to the Chinese language due to its different features from those of Western languages.

The biggest difference is that each Chinese word has a tone, which remains distinctive when grouped with other words to form a sentence and thus makes Chinese sentences sound choppy to Western language speakers. In Mandarin, the official language of China, there are four tones: 1) flat; 2) rising; 3) falling and rising; 4) falling. For speech recognition in Mandarin Chinese, we had to teach the computer to distinguish the four tones. For instance, the computer would have to learn *ma* with the first tone means "mother," with the second tone it means "linen" or "numb," with the third tone it means "horse," and with the fourth tone it means "scolding." Confusing these four tones would cause serious misunderstandings.

Another special feature of the Chinese language is that a word may or may not be a complete unit of meaning by itself and frequently needs to be combined with another word or two other words to carry a meaning. It is crucial to understand how words form units of meaning in a sentence or even just a phrase like *chang chun yao dian*. The word *chun,* which means "spring" by itself, here should go with the *chang* preceding it to make a name. The words *yao* and *dian* are supposed to work together to mean "pharmacy." If the computer accidentally groups the word *chun* with *yao*, it will refer to "aphrodisiac" and distort the meaning of the entire phrase.

As challenging as our tasks were, we overcame all the difficulties. I had very bright team members. Two of them, Bob Di and Zheng Chen, were especially quick learners. They didn't study speech recognition for their Ph.Ds at Tsinghua University, but they soon absorbed my lectures and gained more knowledge of speech recognition from other resources provided by the Microsoft headquarters. They became project leads within two years.

Jianfeng Gao was another successful team member, but he actually fell behind at first. He was initially slow when it came to problem solving. But I saw his potential and encouraged him to ask me questions as well as get help from his peers. Then he did consult Bob and Zheng, who kindly gave him pointers. Jianfeng gradually made progress. He was promoted to be a project lead in his fourth year with Microsoft Research China. Later he transferred to the headquarters. He is a researcher there today.

Microsoft Research China really has helped young Chinese scientists advance as I initially planned. It also has presented impressive research results to the headquarters.

Some of the most amazing achievements came from my team. Within a year, my team developed a four-tone-recognition system, an accurate data entry method, a Chinese dictation device, and a multi-functional statistic language model.

## Presenting Results to Bill Gates

I made a wish on the first day of Microsoft Research China that someday I would walk into Bill Gates' office to present the achievements of the research center and win his applause.

I shared the prospect with my staff, and then it became our common dream. We learned from various sources how challenging it was to present to Gates. At these so-called "Bill G Reviews," Gates would instantly find something inadequate in a report and question the reporting person relentlessly. A Microsoft director couldn't go to Gates without a huge amount of preparation.

I knew product teams of Microsoft generally presented to Gates once a year. But he cared about Microsoft Research more, so the research teams presented to him every quarter.

In June 1999, Rick Rashid came to a conference in Beijing and saw initial results of Microsoft Research China. He said, "Kai-Fu, I was planning to have you report to Bill next February, but it looks that you've already reached the level he would be happy to see. Why don't you go to him in October? I'll arrange for that."

I was pleasantly surprised. Then I told all my staff, "We have to work harder on our projects so we'll have a perfect presentation."

We worked day and night until my October trip to the headquarters. We hardly slept but didn't feel tired because we were totally wound up.

My team bought a lot of text corpora from the People's Daily, China's largest newspaper, to train our Chinese language model. We also merged the model to a larger system, using resources from the headquarters. We had to make the four-tone recognition of our system problem-free before demonstrating it to Gates.

The Multimodal User Interface team perfected their model-less user interface by continuing to fine tune it until the last minute before it was taken to America.

The Internet Multimedia team submitted their results of multimedia compression. A team member Shipeng Lee developed an important part of the international MPEG-4 standard. We wanted to try to get that accepted before the review.

The Image Computing team enabled a computer to present a three-dimensional environment and allowed users to have their virtual avatars do everything in it. The images were even bigger and more lifelike than Apple's QuickTime VR.

The Multimedia Computing team had completed an intelligent image search system.

Once a face in a picture was selected, the system could spot the same person's face in

all the other pictures in the database. The system was also able to categorize video images. For instance, it would put football and basketball in different categories.

On October 18, 1999, I walked into Microsoft headquarters with five other project leads from Microsoft Research China. We were all wearing black cotton jackets to look like a team. I also had black polo shirt on.

Gates' office was in Building 8, which was two-story, with white walls and hunter green window frames. We went to the second floor and entered a conference room, which was very simply furnished, without any luxury items people might expect to see in the richest man's office building.

Gates showed up a little after 10 a.m. in a brown shirt without a tie. He looked very serious. Without using greeting words, he simply nodded to signal me to start.

When I presented to him, he listened attentively and asked questions like "Do local universities think Microsoft is taking talents away from them?" He told me to keep good relations with Chinese universities. "It's as important as the scientific research," he said.

I told him that we wouldn't proactively recruit senior faculty members from universities or research centers in China, but if they applied for our positions, we would seriously consider their applications.

Gates showed a little humor when I told him one of our project leads, Jin Lee, had been a gifted child and once had a chance in his elementary school to meet China's former leader Deng Xiao Ping (Deng being the family name as traditional translation would keep the Chinese way of putting family name first). Deng patted Jin's head and said, "China's computer science should start with kids!"

"I hope you didn't hire him only because Deng had patted his head!" said Gates.

We all laughed. Then I explained to Gates our difference from other Microsoft Research centers. He showed a lot of interest in our mission to create technology applicable to consumer products, and he took notes. He agreed with me that the research center should let every researcher choose work in his or her strongest areas.

When I introduced our Chinese speech recognition system, I was surprised to realize Gates already had prior knowledge of all the difficulties with Chinese data entry. I told him that doubling the speed of Chinese data entry would help Chinese computer users save a billion hours in total for every two working hours. He humorously replied, "You're saving more time than improving Windows boot time."

Last but not least, I talked about our new research directions, namely the next-generation multimedia, user interface, and message processing. Gates responded, "Microsoft Research has succeeded in audio. We'll count on you guys for video and multimedia."

I immediately said, "Bill, three of our teams have achieved many results in multimedia!"

Gates looked very pleased to hear about our multimedia technologies and kept taking notes. Later he often told people, "I bet you guys don't know our best multimedia scientists are at Microsoft Research China!"

His meeting with us was supposed to last only an hour, at the end of which Rick asked him whether we should stop. He said, "No, I want to hear more. I still have 45 more minutes."

When Gates concluded the meeting, he said, "It was an outstanding presentation. Perfect!"

As we walked out of the conference room, Gates patted my shoulder and said, "Aren't you happy working with smart people all day?"

Later I presented to him so many more times that I lost count. But that first time was the most unforgettable.

The next day I took the reporting team from Microsoft Research China out to celebrate our successful presentation. We cruised on Lake Washington. While reveling in the spectacular scenery, I opened an ornate wooden box and offered the Cuban cigars to my colleagues. We basked in the glory from the presentation, and enjoyed the beautiful evening.

## The Hottest Research Center & the Saddest Departure

Bill Gates' recognition boosted our morale at Microsoft Research China. Within two years, we published 28 papers in prestigious international periodicals, did 11 keynote presentations at international conferences and submitted 49 patent applications. On Nov. 11, 2001, Microsoft promoted the center to be Microsoft Research Asia.

While our staff only accounted for 1/30 of the scientists in the related fields in China, our high-quality papers were 170 times theirs. The research center was named the "Hottest Computer Science Lab" by *MIT Technology Review*.

Years later, people still talk about the significance of the research center. One analyst said, "Let's recall in the 1990s Microsoft was accused of its monopoly and lack of willingness to follow Chinese regulations in China. The founding of Microsoft Research China in 1998 was a turning point. In the following decade the research center improved the image of Microsoft in China by recruiting talents and creating new technologies. Bill Gates has once told the media that Microsoft Research Asia has brought a pleasant surprise to the company's brand building in China."

Our success had a lot to do with our unique system. Unlike other Microsoft Research centers, which let every researcher work autonomously, we had project leads assign duties to new employees while training them to gradually take initiatives. This worked out effectively, given our junior staff's lack of experience. But this approach also leveraged their hunger and their perseverance.

Several years later, when I founded Google China, I let each engineer choose projects to work independently like American researchers because by that time Chinese scientists had reached international levels. It was my pleasure to have experienced and contributed to the maturing of young Chinese engineers.

I was also pleased to see some of our research center's innovations becoming Microsoft products, which outnumbered those coming out of the other five Microsoft Research centers, given our applications-oriented approach.

In summer 2000, I received a phone call from Microsoft headquarters. The caller was a senior vice president, Bob Muglia. He said, "Kai-Fu, you've been promoted as a vice president. Steve Ballmer recommended you and Bill approved. You will be one of the supervisors in our new Dot Net Department, which will be based on a new-generation Internet platform. You'll be in charge of all the user interface!"

This was great news, which meant I would lead a bigger team and embrace more challenges. Of course I accepted the new assignment, but I felt a sense of loss.

To me, Microsoft Research China was like my own baby. It was hard for me to leave.

Rick and I decided to pass my leadership to Ya-Qin, who I knew would make the research center continue to excel.

After Ya-Qin and I discussed all the transition matters, I held a meeting with the entire staff. In the beginning of the meeting, I played a video recording of my boss Rick Rashid's announcement that I would be promoted to a higher position in the headquarters. Then I walked to the microphone to give my last speech as managing director of Microsoft Research China.

I said, "Let me thank my boss for giving me the marvelous opportunity to work with you. Let me thank you for your achievement that makes me proud. Let me thank many young people in China for making me see a bright future. And let me thank Ya-Qin for his strong abilities that will for sure help this center create more miracles! Please don't forget: you are the future of Microsoft and the future of China!"

As I spoke, I recalled every night we had worked overtime together, every time we had discussed issues, played poker and smoked cigars... My eyes watered. Many in the audience also had tears in their eyes. A young intern cried out loud. I walked to him, took off my badge and said, "This is for you as a souvenir." The young man suddenly held me and cried on my shoulder. His tears soaked my shirt.

Rick saw the video recording of my last staff meeting at Microsoft Research China. He sent an email describing it to Gates and Ballmer:

*I started out the meeting by announcing Kai-Fu's departure from Microsoft Research China. The reaction from the team was similar to what I would expect if I had announced a death. Clearly, Kai-Fu did an incredible job in creating his team in China. He inspired not just loyalty but the kind of love and respect that you would only see from a truly respected leader.*

I was touched when I later received a copy of the email, which brought back vivid memories of those challenging but rewarding two years. I felt a deep sense of achievement about the ideas turned into reality, and the mission impossible turned into mission accomplished.

*Passing my leadership of Microsoft Research China to Ya-Qin Zhang*

# CHAPTER 9

# Voiceless in Seattle

For my promotion, I moved from Beijing to Seattle and transferred from a research center to a product department. I needed to switch gears and adapt to the culture of the Microsoft headquarters.

Microsoft in 2000 was the largest software empire in the world. It was famous for being competitive. Numerous Microsoft employees told me they believed Microsoft's most magical moments were when the company started out as number two in a certain field, with an aspiration to destroy the number one. The company would have a "playbook" to beat the competitor. This playbook would span from competitive analyses to outspending the competitor, to "embrace and extend." It would even include ways to rally the troops with a maniacal focus on destroying the competition. Internally, the leaders would chant company slogans such as "Break WordPerfect (the dominant word processing software before Word)!" and "Cut off Netscape's air supply!"

However, as soon as Microsoft became number one in a field, it would slow down, and reallocate resources to areas where it was behind. For instance, after Internet Explorer beat Netscape, Microsoft lowered its investment in browser development. Internet Explorer therefore didn't advance for years until Firefox appeared as a new competitor. Then Microsoft began to sharpen Internet Explorer again.

I soon felt uncomfortable with this company culture. It also somewhat bothered me when I was given guidelines on writing work emails in coded language in order to avoid trouble for the ongoing lawsuit, which threatened to divide the software empire into smaller territories.

To fight the accusation of breaking anti-trust law from the Department of Justice, Microsoft worked with a strong team of lawyers, who composed an email writing guide for the company's senior management team. Since the Department of Justice accused Microsoft of bundling browser software with the Windows operating system, we were not allowed to use the word "bundle" when discussing products in our emails. We were not supposed to frankly state, "Let's bundle this new function with Windows to increase market share." Instead, we had to write, "In order to benefit users, we will introduce integrated innovation by adding this new function to Windows."

Microsoft lawyers paid attention to every little detail to ensure nothing could be used as evidence against the company. In the meantime the company's PR team painted a ignorant image of the U.S. government, claiming the Department of Justice knew too little about high tech to judge what Microsoft was doing.

Bill Gates insisted that Microsoft had done no wrong. He was personally involved with the legal battle. He submitted a 155-page testimony to convince the court that Microsoft kept consumers, the prospect of personal computers and the future of American economy in mind despite malicious attacks from competitors.

In 2001, Microsoft and the Department of Justice settled out of court. Gates definitely deserved credit for saving the company from being divided. But by that time he had decided to leave his CEO position for spending more time on philanthropy and with his family. He worked as chief architect and chairman of the board after passing the CEO position to Steve Ballmer.

According to some Microsoft employees, Ballmer was once widely noticed in 1994 at a company meeting. Many employees saw him dance around the stage, heard him shout, "Market share! Market share!" and watched him gesture choking opponents during his performance.

Ballmer took over as a different type of CEO than Gates, who would shut himself in a waterfront cabin for a "Think Week" every year to draw new blueprints for the Microsoft empire. Ballmer left technology to Gates and put his own focus on the company's operations.

Ballmer was well recognized for his morale boosting techniques. At the annual Microsoft sales conference in New Orleans in 1998, Ballmer yelled at the top of his naturally loud voice, "I love Microsoft! I love Microsoft! I love Microsoft!" He inspired 50,000 sales team members to cheer for longer than five minutes!

However, when Ballmer became CEO, Microsoft had low morale due to the impact of the lawsuit on the company's reputation. He also had to deal with dot com bust, which made it difficult to keep Microsoft's stock prices from sliding, especially given no new products at the time.

Ballmer attempted to change Microsoft's image through internal company slogans such as "Open but respectful," "Integrity and honesty," and "Let's put users first!" But Microsoft's action didn't seem to match these words.

In summer 2001, Microsoft launched a new policy called "licensing 6.0," which forced every user to pay a maintenance fee. The maintenance fee was higher than the upgrading fee, and those who didn't pay the maintenance fee would have to pay the full price of the new version when upgrading the software products they had bought from Microsoft.

The policy came out during an economic downturn, so many users complained about the twisted way of raising prices. It essentially broke Microsoft's promise to old customers. Numerous senior staff members of Microsoft felt terrible about it. But no one challenged Ballmer until June 2002.

At Microsoft's annual meeting of global vice presidents, Ballmer gave a speech on new company values. Just when he explained how to put users first, Orlando Ayala, group vice president of Global Sales and virtually the number-three person of Microsoft, stood up.

"Steve," Alaya looked right into Ballmer's eyes. "Today I feel ashamed of being a Microsoft person."

Everyone's eyes opened wide.

"Because of our lawsuit, my children suffered from gossip in school. Their classmates said their father broke the law and denied it," Alaya said. "That was OK, as long as we can truly prevail in the lawsuit. But licensing 6.0 makes me feel hopelessly ashamed. And now all the slogans about honesty and integrity and customers first just make me feel like a hypocrite!"

"Our new policy is forcing users to pay us more," he continued. "But you are asking me to tell them we put users first at the same time. How can my conscience allow me to do the job?"

"We all joined this company to change the world with cutting edge technology. But this new policy is telling me that our values consist in making money and beating competitors," he concluded. "Sorry, Steve! I can't continue to lie to our users!"

Tears streamed down his cheeks.

I felt as if he had spoken for me, saying everything I wanted to say. All the other vice presidents at the meeting seemed to feel the same. Someone started to clap. Soon everyone was applauding for Alaya.

Ballmer's eyes were filled with anger. But apparently he knew this was not a time to let it out. He simply said, "OK, Orlando, we heard what you had to say. I'll talk to Bill tonight about it and give you an answer tomorrow. Now, let's move on to the next item on the agenda."

## Gates' Tears

The air was still, but full of tension. The huge conference room was so quiet that a pin drop could have been heard. All the Microsoft vice presidents were waiting in silence, wondering how Bill Gates and Steve Ballmer would respond to Orlando Alaya's penetrating comments.

In the beginning of the meeting, Gates said, "I'd like to take an hour of your time. Let's move all the items back, OK?"

We knew his answer was about to begin.

"Did you know why I created Microsoft?" Gates opened up with a rhetorical question. "It was not for any sort of competition. I had a simple objective: I wanted to put a computer on every desk. To realize this dream, I knew computer prices would have to be lowered first. And I knew only standardizing software could lower computer prices. That was why I founded Microsoft. So, who can say Microsoft is not putting users first? I'd be the first one to object to that!"

"Did you know why I'm still working here?" Gates continued. "Many of you don't know my family lives in a fish bowl, being watched by the whole world. Do you know how painful it is? My children receive threats from kidnappers every month. Just two years ago, I received an anonymous email with pictures of my daughter's hourly activities attached. The guy said, 'Mr. Gates, how would you feel if what I pointed at your kid was not a camera but a rifle?' A week later, he asked me for 10 million dollars. I paid him and then called the police. The police caught him, but guess what? He didn't serve much time in prison because of no real criminal act and no prior criminal records. Actually he was released yesterday. Did you know my wife was unable to fall asleep after hearing the news?"

"It's so painful to live the public life here at Microsoft. Why am I still here?" Gates paused a few seconds and then firmly said, "It's because I have to fight those who call us a selfish monopoly that takes advantage of users! Why don't they investigate who has created so many job opportunities in software and benefited so many users? Why don't they try to understand what kind of person I really am? Why call me an arrogant monopolist? Why blame me for taking advantage of others? For our industry, our users, and our company, I've sacrificed my private life and my family. Why are you doing this to me?"

He became too emotional to finish his speech. Tears flooded his eyes. Ballmer walked onto to the stage to hug him.

No one knew what to say any more. It was shrewd of Ballmer to bring Gates out to confront Alaya, because all the vice presidents liked and revered Gates as a technically brilliant leader with a down-to-earth personality, always friendly with everyone.

No one wanted to make Gates feel worse, so Alaya didn't get any more support in public. However, numerous vice presidents, including me, still believed Alaya was right. We privately expressed our admiration for his honesty and courage.

A few weeks after the meeting, Alaya was transferred out of Global Sales and demoted. What happened to him cast doubts about Microsoft on my mind. While I still felt attached to Microsoft for the great opportunities it had given me, I began to wonder about the company's directions, especially when looking back on my first project at the headquarters.

## The Aborted .Net

In July 2000, I began to take charge of the user interface of Microsoft's .Net Group, which was a new department. Bill Gates decided to establish it in order to transplant the Windows platform and software to the Internet.

The .Net Group was assigned to create an on-line Windows system, which would incorporate wireless communications, intelligent appliances, and new user interface. As I familiarized myself with my new job, I realized the biggest challenge was in team building.

I was not taking over an existing department or recruiting talents for a new group. The company was not growing. I had to identify groups that I wanted, and then work to get them re-organized to my group. Needless to say, it was exceptionally difficult to take team members away from their vice presidents!

Another problem was a disagreement in Microsoft's top management. One member of Microsoft's Senior Leadership Team, Jim Allchin, vehemently opposed the idea of transforming Windows. Allchin had earned a lot of credit for overseeing the creation of Windows 98 and Windows XP. He often said, "My blood is in four colors (like the Windows logo)." He saw Windows as Microsoft's core technology, a fundamental change of which would be inconceivable to him.

In summer 2000, Allchin was on sabbatical. Paul Maritz and a few other vice presidents who saw the increasing importance of browsers seized the opportunity to suggest transplanting Windows functions to a browser. The prospect they presented to Gates was very similar to the cloud computing Google and IBM are doing today, except for being limited to a single platform. These Microsoft vice presidents were planning on creating a super browser, which would be very similar to Google's Chrome OS today. It is fair to say that Maritz's vision was at least five years ahead of even Google.

Gates approved the proposal in Allchin's absence. Then the .Net Group was formed. But Maritz, who was supposed to lead the new project, took his sabbatical, so all the organizational responsibilities of the group fell on Bob Muglia, the senior vice president whose phone call had brought me to the headquarters, and his management team of three members, including me.

Gates was quite enthusiastic about .Net. Given my strong technical background, he often asked me to discuss the new project with him.

At a one-on-one meeting with Gates, I described my ideal blueprints of the project, "Bill, I think the super browser should have four parts. First, all the browser experts of Microsoft should work together to create a browser platform where all the application software will be usable, so there will be no difference between application software and a website. Secondly, we should transform MSN into cloud computing to include functions such as email, instant messaging and login. Third, we will transplant MS Office to the Internet so users will have a permanent cloud-based storage, and never lose their files. Then we can charge for extra services such as extra storage, high-quality printing and machine translation. Last but not least, we should develop new programming technologies to combine the functions of the Internet and the front end."

That was a memorable time of my career. Gates repeatedly said he was happy about having me in the headquarters. A New York Times article told the story of my promotion when mentioning Gates' "new inner circle" of seven executives (interestingly, the article was also written by John Markoff, the reporter who had written about my Ph. D. thesis). As depicted in the article, Gates valued my opinions and gave me a lot of encouragement.

However, Gates was not as unchallengeably authoritative as many might think. His support didn't necessarily make a project a must-do to everyone, especially if it involved relocation of resources.

Based on the plan approved by Gates, I would integrate three teams: Internet Explorer, MSN Explorer and NetDocs. It was much easier to combine the technologies of the three teams than to bring their people together.

MSN Explorer was a team that had spun off from Internet Explorer, so the two teams held a grudge against each other. NetDocs and MSN Explorer didn't get along, either, for some technical issues. Another team, Office, hated NetDocs even more because the mission of NetDocs was to keep files on line and that would replace the usage of Office. For the same reason, Office also saw .Net as an enemy. I certainly couldn't expect anything from Office and didn't even try.

I was only able to get MSN Explorer to move under me. It was a strong team but had merely 100 people. And we couldn't start the .Net project right away because the team members of MSN Explorer had to finish their existing tasks first. I continued to talk to NetDocs and the IE teams, but progress was slow.

One day, someone told me, "Uh-Oh! Allchin is coming back soon!" Apparently many Microsoft employees were afraid of Allchin's tough personality. They predicted that Allchin would go ballistic when he learned Windows was going to be transformed. And they were right.

Allchin yelled at us, "Do you know what's paying your salaries? It's Windows! My blood is in four colors. What about you? Are you guys cold-blooded?"

Then Allchin went to Gates, threatening to resign if Microsoft insisted on carrying out .Net. He also told Gates that Windows would have a larger database and cooler technology in the near future, so it would be unwise to take it apart now.

At this time Maritz was still on sabbatical. My immediate supervisor Muglia didn't dare to say a word. Gates didn't hear anyone challenging Allchin, and he really didn't want Allchin to leave. So, he changed his mind about .Net.

Gates and Ballmer gathered all the vice presidents at a meeting to announce the cancellation of the .Net project. Gates asked everyone to concentrate on Windows Vista instead.

Maritz never returned from his sabbatical. The .Net Group was dissolved. Microsoft held a meeting for reorganization and asked each vice president to provide input.

When it was my turn, I mustered all my courage to speak my mind, "Our company has more IQs than any other company. But we continue to fight each other, and work against each other. It is a pity that we do and then undo these projects haphazardly. All the IQs that we have become not additive but subtractive!

No one responded to my talk. But after the meeting, numerous colleagues emailed me to say, "Kai-Fu, I wish I could've had your courage." Or "I wish I would have dared to say what you said."

Gates later expressed his understanding of my feelings and agreed that too much political struggle and reorganization was not healthy. He would even use my words to caution employees who kept arguing with one another, "Don't forget Kai-Fu said our IQs became subtractive."

## Working with MSN

After dissolving the .Net Group, Microsoft renamed .Net to be a "brand," which later became just a way of programming. The reorganization of Microsoft put me in a new position in charge of MSN search, as well as corporate speech and natural language interface. On my first day with MSN, someone came to ask me, "Kai-Fu, would you like to meet our new boss, Rick Belluzzo?"

"What?" I was shocked. "Rick Belluzzo?"

My old boss at SGI! The one who gave me no choice but to sell my department! I couldn't believe I was going to work under his supervision again!

Belluzzo lost hope in SGI and left his CEO position for Microsoft, so he became my immediate supervisor again. I felt as if being tricked by destiny. When Belluzzo saw me, he looked surprised, too. But he smiled and said, "Kai-Fu, let's forget everything unpleasant in the past and focus on the future."

I nodded, and then returned to my desk to immerse myself in my new job. Gates had told me that he wanted the speech and natural language technologies to be combined in order to build a more powerful natural language engine that could go beyond search.

He said, "Instead of just finding many documents, can it just do it for me?"

I knew exactly what he meant, and I wanted to make the same thing happen. For example, when I wanted to buy flowers for my mother's birthday, why would I have to search for flowers, and go through all the trouble typing my credit card number, her address, etc.? I asked myself, "Can there be an intelligent assistant to whom I'll just say 'send flowers to mom on her birthday,' and it gets done?"

Gates called this "universal type-in line." He hoped that my expertise in speech and natural language, combined with all the Microsoft experts who worked on speech and natural language, could make this dream come true.

But after understanding his vision, and examining what Microsoft had, I realized that the dream was unrealizable in the short-term (though Gates continued to try, with efforts including the 2009 launch of the "decision engine" Bing). I also saw that the teams re-organized to me were mixed in quality, with some groups going in completely wrong directions.

At the same time, people were searching more and more on the Internet, and Google's search engine was taking off. MSN didn't have any Internet search technology in this important sector. Microsoft contracted search out to Inktomi in the United States and Alta Vista in Asia. I felt that instead of going after the science fiction dream of "universal type-in line," we needed to address this immediate market need – developing a search engine. I saw two options for MSN: 1) working on creating a search engine; 2) buying a company with a good search engine. I played with Google quite a bit, and became very fond of it myself, so I recommended the acquisition path to Gates.

Gates once considered my suggestion of buying Google. But after initial negotiations, he found Google's market value exceeding 500 million dollars, and to buy Google would require at least one billion dollars.

"It's a company without revenue but asking for a billion dollars. Those two kids (Larry Page and Sergey Brin) are crazy!" Such a comment was Microsoft top management's typical reaction to the talk with Google.

Little did they know! It was probably beyond their imagination that Google would be worth 200 billion dollars six years later.

After dropping the idea of buying Google, Microsoft was not in a hurry to create its own search engine because search was deemed low value at the time. Steve Ballmer felt that Microsoft's competitors were Yahoo! and AOL. Google wasn't seen as a potential threat at all.

Gates told me Microsoft would put search on hold and would have me join the company's most important project, Windows Vista, so I left MSN in early 2002. But a year later, to my surprise, Gates approved MSN to create a search engine in February 2003 and hired Christopher Payne, formerly sales vice president of Amazon, to take charge of it.

MSN in 2003 promised to catch up with Google within a year. But it didn't launch a domestic search service until 2005. According to comScore statistics, Microsoft's market share in search was 15% in October 2005 and then dropped continuously to only 8.3% in January 2009. During those few years Microsoft provided gifts for its search users and promoted its search in IE 7.0. In 2008, Microsoft even offered 44.6 million dollars to buy Yahoo!, hoping to combine its own search market share with Yahoo's 23% to fight Google. However, the deal unexpectedly fell through after Yahoo! co-founder Jerry Yang declined the offer.

On June 4, 2009, Microsoft launched a global search engine and called it Bing. By that time Google's American market share had reached 67.5%. Microsoft was a decade behind Google on the development of the search business.

## Developing Windows Vista

When I joined the Windows Vista Group in 2002, I reviewed the history of Windows. I remembered Bill Gates being interviewed about Windows 95 by Jay Leno. Gates humorously said, "Windows 95 is so easy to use that I'm sure a talk show host can operate it, too."

Windows 95 was a milestone in the history of technology. It gave the user the first complete operating system with graphics and made working with computers interesting.

Windows 95 came from a tremendous amount of collective effort. Tens of thousands of Microsoft engineers worked on it tirelessly. A product manager told me, "In the process of developing Windows 95, we consumed 2,283,600 cups of coffee and 4,850 pounds of pop corn."

Microsoft kept investing more in Windows to stay ahead. After Windows 95, the company launched Windows 98, Windows 2000 and Windows XP. Windows Vista was one more attempt to bring Windows to the next level.

Gates had me create a new division called Natural Interactive Service in the Windows Vista Group. We set three objectives:

- Supporting C# language, which runs slowly but develops fast. This would help Microsoft keep up with Linux, which was developed by an enormous number of volunteers all over the world.

- Developing Windows File System (WinFS), which could save all kinds of files in its database. If it succeeded, all the data of the world would eventually go into Microsoft's database, and Microsoft would beat competitors in database such as Oracle and IBM, plus preventing Google from taking control of more data.

- Developing Avalon, a next-generation display system that would make traditional software look the same as a website in the browser.

Meeting the three objectives were all extremely difficult. One director complained, "C# is so slow. How can we use it to create an operating system?"

"Our database is also not fast enough. How can we use it as a file system?" added another.

We were taking a leap of faith to work towards the three objectives. In the meantime we came up with a few exciting new functions:

- Windows Intelligent Assistant, which would come out at a click and might even automatically come out to solve problems for users.

- Executive Assistant, which could execute the user's requests such as printing and email forwarding.

- Machine Learning, making the computer absorb new knowledge on its own.

- New File Processing, which would find any file searched by the user in any database and convert the file into part of the WinFS.

From 2002 to 2004, the teams under my supervision were entirely dedicated to the development of Windows Vista. But it looked that we were understaffed. I looked for the causes of the problem and found a 150-member team doing useless work. Half of the team members were linguists who didn't know anything about technology. The team leader was counting on the linguists to form a "language rainbow" that would help computers understand human language more deeply.

I knew his idea was not feasible. I was as sure about that as I had been about dismissing the expert system from my Ph. D. research on speech recognition. I decided to lay off the linguists and transfer the team's engineers to other teams to do promising projects.

When I informed the team leader of my decision, he refused to cooperate. Instead he took the case to Ballmer, who in turn brought it to Gates.

Gates asked, "Kai-Fu, it's a good idea to enable machines to naturally communicate with people. Why do you insist on eliminating the Natural Speech Processing Team?"

"Because the team is in the wrong direction," I explained.

"But David is an expert, too," Gates said. "He disagrees with you."

"Bill," I reminded Gates. "David is an expert in the operating system. I'm more of a speech and natural language expert."

"But we've invested a lot in the project," Gates hesitated.

"Bill, when you are in a wrong direction, the more you invest, the more you lose," I provided my analysis. "Then it may eventually be too late to make it up."

I looked into Gates' eyes and sincerely continued, "Many people in this company will say many things to you for their own benefits. But I promise you I will never lie to you."

In that instant, I felt our souls touching each other.

"OK," Gates nodded. "Let's do as you suggest."

## Rewriting Windows Vista

Steve Ballmer once said, "At Microsoft, the only thing that doesn't change is change!"

Given his description of Microsoft, it shouldn't have been a surprise that the launch of Windows Vista kept being postponed. However, many team members still felt frustrated. They were afraid of never being able to overcome the technical difficulties in WinFS. The morale was incredibly low.

Finally, top management recognized postponing couldn't solve the real problem. In fall 2004, the big boss of Windows Vista, Jim Allchin, held a meeting with all the vice presidents in the group, including me.

Allchin said, "We've confirmed that we cannot complete Windows Vista on schedule. Based on our current progress, we can't foresee the launch date, either. So, we have no choice but to redesign the product. I'd like to ask you guys, if we start over in the hope of finishing it in two years, how much do you think we can do?"

We were all shocked to hear his announcement. Everyone took a deep breath, and looked at one another as if asking, "Who's responsible for this?" "What does Bill think?"

Allchin continued, "I know, you must all want to know how Bill responds to this. I first told him about it two weeks ago. He didn't believe it at first. But after consulting a number of technical executives, he held a meeting with me last night. He said he'll just let us decide what to do."

While continuing to explain the decision to us, Allchin didn't say he was going to take responsibility for the failure. It wasn't until days after the meeting that he announced his early retirement.

At the meeting, all the Windows Vista vice presidents discussed future options and reached an agreement on: 1) stopping the use of C#; 2) Abandoning WinFS; 3) Revising Avalon.

Half a day later, I went back to the nearly 500 people under my supervision and told them the bad news. I said, "I know it may sound really difficult for you to accept. But if we keep following our original plans, we really can't complete Vista. So, it's a wise choice to start over. I hope you won't let the change bother you too much. Let's cheer ourselves up and move on!"

All my subordinates looked disturbed, but none of them said a word. I knew they must have felt what I felt. As explorers in science, we had to stop going toward a dead end and start taking another direction regardless of our hurt feelings.

We pulled ourselves together to determine what to keep and what to dump out of what we had done. Unfortunately, we realized we had to throw away most of our tasks, to which we had dedicated two and a half intensive years!

We were more disappointed at what the new guidelines left us to do. It didn't look that we were able to contribute much.

In the end, when Windows Vista was finally launched, two of its 10 major functions came from my division. That sounded pretty good, but it only accounted for 10% of our original creativity.

## Bridging Microsoft with China

After I left Microsoft Research China for my promotion, my job duties were completely unrelated to China. But given my successful experience there, top executives at Microsoft often consulted me on China issues, especially when they faced problems in the China market.

While I was busy with Windows Vista, Microsoft made a serious mistake in China. After promising the Chinese government to give China $100 million worth of business in the next three years, the Microsoft executive in charge of the contract negotiations claimed the money was going to Microsoft Research Asia and Microsoft's R & D Center in China.

Chinese media sharply questioned the announcement. A journalist wrote, "How can funding Microsoft's China branches count as giving business to China?" Another published an article titled, "Why Is Microsoft Playing Games?"

Microsoft CEO Steve Ballmer complained to me, "What's going on? When I signed the contract with the Chinese government, they (some

Microsoft executives) told me the 100 million would go to Microsoft Research Asia and our China R & D. Why has it become such a problem?"

I shook my head, "They shouldn't have told you that. Let's think about it. How can we say funding our own China branches is giving China business? Do you think the Chinese government can take that? We have to solve this problem soon. Or Microsoft's image in China would be totally ruined!"

Ballmer realized the severe impact of the incident and responded furiously, "Who was the idiot that put together the contract? I'm going to fire him!"

"No matter who's responsible, we need to solve the problem first," I provided my advice. "What about moving $100 million worth of outsourced projects from India to China? If I can find suitable Chinese companies, would you be willing to do that?"

"OK," Ballmer nodded. "I guess this is the best solution. I hope you can find some Chinese companies capable of doing what those Indian companies do."

Before carrying out the assignment I had brought upon myself, I needed to appease those managers in charge of outsourcing to India. I gathered all of them, explained the situation and said, "To solve our crisis in China, we must give China $100 million worth of outsourced projects. So, we have to take some outsourced projects away from India. For the ultimate good of our company, please understand!"

With all the possible road blocks cleared out of the way, I went ahead to work with then-Microsoft-global-business-development-manager David Han in search of potential outsourcing partners in China. We found six Chinese companies that seemed able to offer what Microsoft needed. Then I asked a Microsoft testing manager, Charles Clark, to cooperate with David and take charge of negotiations with the six companies.

Charles was happy about the assignment partially because he wanted to see his Chinese wife's homeland. He and David visited the six Chinese companies to train each on how to do business with Microsoft.

After three weeks of intensive training, each of the six companies sent two representatives who were communicative in English to the Microsoft headquarters to negotiate contracts. They were careful about only asking for projects they were able to do well because they knew the first deal would make or break their future prospect with Microsoft.

The six companies helped Microsoft keep its $100-million promise to the Chinese government. All the Chinese bad press about the incident disappeared. Ballmer was pleased. Then he asked for my advice on the China market more and more frequently. He knew I had no stake in China after transferring to the headquarters, so he believed I would give him objective opinions without ulterior motives.

In order for Microsoft's top management to understand China better, I organized a China tour for 20 vice presidents of Microsoft. I took them to the Imperial Palace in Beijing, where they watched a costumed show about a Chinese emperor, Qianglong (1711-1799), during whose reign China was a strong country. I told them more historical stories, and I let them experience contemporary China through shopping. I asked each of them to buy an outfit for less than 200 yuans (about $25) and see who would be able to spend least. Then they realized prices were really low in China. They finally understood the fixed price tags on Microsoft products worldwide could be unaffordable to many Chinese.

I also brought the vice presidents to see Chinese government officials, who were in general friendly and courteous, very different from the hard-core stereotype. Some of the younger government officials spoke fluent English to the Microsoft vice presidents and displayed a surprisingly progressive attitude.

At the end of the tour, the Microsoft vice presidents met Chinese college students at a banquet and saw China's future hopes.

When we returned to the Microsoft headquarters, the vice presidents told me that the trip had erased a lot of their bias against China. While pirating in China still looked very bad to them, they saw the core of the problem in the developing country where people were too eager for money to respect copyright, and they recognized China's effort to move up in every way.

In order to give Microsoft an even deeper understanding of China, I wrote a 15,000-word report, "Making It in China," which clearly explains in detail how an international company can succeed in the China market. It was meant to help Microsoft's China policy makers. But later, when some universities invited me to lecture on international companies' development in China, I always drew data from the report to give those speeches.

## Putting Out a Fire for Bill Gates

In March 2003, Steve Ballmer assigned me to accompany Bill Gates to visit China. The assignment included a mission for me to understand and solve the many problems Microsoft was facing in China during that time.

The problems had been looming for quite a long time. While Microsoft Research Asia kept making breakthroughs in technology, the China branch of Microsoft Corporation didn't do so well in sales. In late 2001, the city government of Beijing made large purchases of software and awarded contracts to a Linux vendor as well as several Chinese software companies, but didn't order anything from Microsoft. One of the major reasons was Microsoft's negative image in China.

Chinese media often criticized Microsoft for not respecting Chinese law and not understanding the local culture. They also said there was a "Microsoft curse," which made all the Microsoft China presidents leave within two years.

Juliet Wu was one of the former Microsoft China presidents who did leave in less than two years' time. She joined Microsoft in February 1998 and resigned at the end of 1999. Around then she published a book titled, *Flying against the Wind*, which sharply critiques Microsoft. The book instantly became a best seller in China and did a fair amount of damage to Microsoft's image there.

Wu's successor, Jack Gao, worked as Microsoft China president for about 27 months, only a little longer than two years. After he left in March 2002, he also published a book, titled *My Microsoft Experience*, which describes Microsoft's problems in detail.

According to the book, when Gao held his first staff meeting at Microsoft China, nine out of ten questions brought up were about why the company was so terrible at public relations and government relations in China.

It was during Gao's reign that Microsoft China completely lost the deal with Beijing City to Linux and local Office suites. This was a disaster because Beijing was not only a large customer, but it also set the tone for the rest of the country. Gao explains in his book that the biggest reason for the failure was his disagreement with president of Microsoft Greater China. He says, "The Beijing City case was assigned to Microsoft Greater China. Microsoft China was only supposed to provide assistance. Microsoft Greater China refused to lower price quotes and made strict demands [to the city government of Beijing]. That's why the purchases of operating systems and office software didn't even include one single Microsoft product."

Gao comments in his book, "Microsoft Greater China and Microsoft China have never had successful cooperation. Is it because the executives of the two branches don't get along? Or is it because the system itself is unreasonable? It's hard to pinpoint where the problem exactly is. But the mechanism of the system can cause gaps between executives [of the two branches], who may create insurmountable obstacles against each other."

Since the system put Microsoft China under the management of Microsoft Greater China, Microsoft China president often felt restricted. The situation remained the same after Jun Tang succeeded Gao.

Numerous staff members of Microsoft China were unhappy about certain policies set by Microsoft Greater China, and they emailed the headquarters to express their discontent. I frequently received their lengthy complaints. They even reached Ballmer sometimes.

Microsoft Greater China did make one mistake after another. After losing business with the city government of Beijing, Microsoft Greater China worsened the company's government relations by inviting some anti-Chinese-government activists to give speeches.

To improve Microsoft's relationship with the Chinese government was one of the major goals Gates planned to achieve during his visit. After we arrived in Beijing and checked into our hotel, I gave him some advice in private.

"Bill, the Chinese government is quite unhappy with Microsoft these days," I said. "Should you hear anyone saying anything unfriendly, would you please ignore it? Please don't talk back in that case. I'll do my best to solve all the problems for you, OK?"

Gates looked at me and smiled like a child. "OK, no problem," he said. Then he shook his head and added, "Ah! Now I see you and Steve have tough jobs."

When I walked away, I heard him murmuring to himself, "Um! I'm glad I no longer have to deal with this!"

During his three-day visit, Gates first went to see China's president Jiang Zemin (Jiang being his family name as the official translation keeps the Chinese way of putting family name first). Later Gates met with Chinese scholars, educators, researchers and college students.

He also introduced Microsoft's .Net technology to representatives of more than 600 Chinese software companies.

All the events went smoothly. Gates did take my advice to heart. Whenever facing sharp questioning, he maintained his gentlemanly attitude and said, "We'll reflect upon our mistakes" or "We apologize."

In the meantime, I did everything possible to ensure harmonious communication between Gates and everyone he met. Before a meeting with a Chinese government official who had openly criticized Microsoft, I went to see the official and expressed Microsoft's sincerity in solving all the problems in China. The government official finally promised me not to give Gates a hard time, and I promised him I would ask top management of Microsoft to look into Microsoft Greater China's mistakes.

When I walked out of the official's office, I felt relieved. But suddenly I realized it was time for me to accompany Gates at a lecture, and I was supposed to speak after his speech! I immediately asked my driver to take me to the university where the lecture was taking place. Fortunately, the 8,000 students were just clapping for Gates' speech when I got out of the car. I arrived right on time!

After Gates returned to the headquarters, I stayed in China to investigate all the problems with Microsoft China. I talked with Microsoft China employees and visited Microsoft's partners in China.

I submitted a longer-than-20-page report on Microsoft China to Ballmer when I was back in the headquarters. In the report, I made four suggestions:

1) Combining president of Microsoft China and president of Microsoft Greater China into one position, China CEO, to take charge of all Microsoft's business in mainland China, Taiwan, Hong Kong and Macau.

2) Making China CEO as high-level as a corporate vice president in order to attract the best candidates for the position.

3)    Letting China CEO directly report to the headquarters instead of going through president of Microsoft Asia.

4)    Helping all the top executives of the headquarters understand special features of the China market and how to succeed there.

Ballmer accepted my suggestions. Microsoft hired America's largest head hunting company to seek an ideal candidate for the new China CEO position worldwide. In August 2003, Tim Chen took office as Microsoft's China CEO and corporate vice president. Chen was indeed the right person for the job. He helped Microsoft improve relations with the Chinese government and increase sales in China.

As I once explained to Microsoft's top management, it requires good government relations to run an international business smoothly in China, where the government not only is a major client and a policy maker but also owns all the media. An international company on the good side of the Chinese government will receive detailed guidance on how to work with Chinese law and establish friendly connections with the local media. In contrast, an international company that has offended the Chinese government will face many obstacles in China. It's unwise of an international company to take an opponent's stand against the Chinese government.

I was glad Microsoft finally improved its relationship with the Chinese government. However, while Microsoft China was doing better, a big problem happened to Microsoft Research Asia. After my successor Ya-Qin Zhang transferred to the headquarters, Microsoft Research Asia seemed to have lost directions. Sometimes two of its departments would be competing for a talented student who had passed an interview. It was also inefficient that Microsoft Research Asia reported to 18 departments of the headquarters.

In March 2005, I suggested establishing a Microsoft Research and Development Group to ensure cooperation among all the company's R & D departments. Later Ya-Qin took charge of forming the R & D group. This suggestion was my last contribution to Microsoft.

## Leaving Microsoft

Many of those who left Microsoft expressed their mixed feelings, to which I resonated when I was thinking about leaving.

While losing passion for the job I was doing for Microsoft, I still appreciated the company for investing billions of dollars every year in research. The new technologies Microsoft had created were changing the world. Bill Gates' dream, to put a computer on every desk, was coming true.

I admired Gates for asking every product and technology department to report to him in order for him to ensure no overlap between what any two of the departments were doing and to enhance cooperation among all these departments. In the meantime, all the teams thoroughly reviewed their projects during their preparation for their presentations to Gates. This policy certainly helped Microsoft keep itself on the cutting edge.

I saw another secret to Microsoft's success in its recruitment of talents. Microsoft would do everything to win over a distinguished scientist or engineer. One of the most widely circulated stories was how Gates persuaded Jim Allchin to join Microsoft. At first Allchin rejected Gates' invitation and even said, "Microsoft has the worst software in the world. I don't understand why you would ask me to join." Gates not only didn't take offense but also sincerely said, "It's precisely because our software has all kinds of flaws that we need an expert like you to fix them." It was his sincerity and big-heartedness that touched Allchin and won him over.

When I was in the management of Microsoft, I also actively recruited talents for the company. I kept calling Dr. Feng-Hsiung Hsu, the inventor of the deep thought Chess machine. Finally he agreed to join Microsoft in 2003. Although he didn't work for my division, I was still proud of bringing him in.

Microsoft valued talented engineers so much that Steve Ballmer would ask every vice president to provide a report on each division's 50 to 150 best employees. The reports about a total of more than 600 employees would be put together as a book, which Ballmer would read thoroughly. He would also have dinner with many of them by inviting 20 of them at a time. Later the engineers would be blown away when they ran into Ballmer, because he remembered all their names and their conversations.

While rewarding outstanding employees, Microsoft also kept an eye on those falling behind. The company asked every vice president to divide subordinates into four categories: 1) exceeding expectations; 2) meeting expectations; 3) meeting most expectations; 4) not meeting expectations. Those in the last category would receive an ultimatum to improve their performance.

Microsoft was as strict to executives as to engineers. Once Ballmer held a meeting with 100 vice presidents and said, "I'm asking you to find the worst 5% of our staff and make them improve or leave, regardless of their seniorities, which means today we have 100 people at this meeting but probably will only have 95 at the same type of meeting next year."

Everyone took a deep breath upon hearing this announcement. And Ballmer wasn't joking. In fact more than five vice presidents were let go after that meeting.

Microsoft operated like a large machine. Everyone seemed a replaceable part. Although I was considered one of the best executives, I was unable to move the entire machine. My product ideas were rarely considered after I transferred to the headquarters. I realized there was no way for me to maximize my influence at Microsoft.

"Should I leave?" I started asking myself. More and more often, when I asked myself this question, I heard a big YES from the bottom of my heart.

# CHAPTER 10

# Microsoft, Google and Me

I used to be a quiet scientist. Never did I imagine myself in the spotlight for what some media called "lawsuit of the century" until it happened overnight.

"What is it about this Kai-Fu Lee? Why are Microsoft and Google fighting over him?" People curiously asked such questions in the summer of 2005, the most difficult summer of my life. Microsoft sued me. Media followed me. Rumors haunted me. It was all overwhelming, and all of it was because of a supposedly simple job change.

## "We Should Circle Him Like Wolves."

I started thinking about leaving Microsoft in March 2005. I wanted to work in China again, but there were no suitable opportunities at Microsoft's China branches. I was keen on being a part of the phenomenal growth in China. I was writing a book titled *Be Your Personal Best* in Chinese for China's Generation Y. In the meantime I was looking for a job opportunity that would bring me back to China.

Given my familiarity with American corporate culture, my understanding of the Chinese market and government relations, plus my experience working for Microsoft Research China, I believed I would be an ideal candidate for an executive's position in the China branch of an international company. I would bridge the gap of communication between the two cultures.

But I would not work for just any company. Microsoft was a great company. I wanted to find a company even better than Microsoft, a company from which I could learn a lot.  With that thought, Google was the first to come to my mind.

Unlike the long established Microsoft, young Google appeared to operate with tremendous energy. Google in 2005 also seemed enigmatic. Their products were insanely great. But the greatest mystery was: how did such a young company create so many great products so quickly with so few engineers?

I read *The Google Story*, the Google-authorized best seller by David A. Vise and Mark Malseed. I particularly liked the episode about how Google co-founder Sergey Brin decided to keep the company's homepage simple:

Lacking the funds to hire a designer and the artistic talent to create something elegant, Brin kept the Google homepage simple. From the start, Google's clean, pristine look attracted computer users hunting for information. In a cluttered world, its primary colors and white background conveyed purity, with universal appeal.

Everyone knew Google had an outstanding search engine and a working environment as much fun as Disneyland. Those who had visited the Google office building would tell outrageous stories:

"Hey! Did you know a Googler can sit on an exercise ball to write programs?"

"Can you believe they come skateboarding into the office?"

More surprisingly, Google even hired Charlie Ayers, former chef for the Grateful Dead, to cook free meals for employees. Everyone in Silicon Valley wondered how a company could spoil employees so much and be so successful. It sounded incredible and incredibly cool!

Back then, no one but Googlers knew what led to Google's success. Google was very open to every employee but secretive to outsiders. They didn't even want to go public because that would require them to publicize their financial statements, which would reveal how much income their search engine was able to generate, and tip off competitors (like Microsoft).

If a computer engineer left a job for Google during that time, former colleagues would say this person disappeared like being kidnapped by aliens. They never heard from him/her about his/her work at Google. It was a miracle that Google, with over 2,000 people, managed to prevent anyone from leaking any of the company's business secrets.

Google also created a Silicon Valley legend with 1,000-fold return to its VC investors. Silicon Valley's VC returns are above stock market only because of Google; if Google returns are removed, then the VC returns are no better than the stock market.

Many were surprised to find out the legend was created by two graduate students named Larry Page and Sergey Brin. *The Google Story* describes how the two founders of Google met at Stanford University as follows:

*When Larry met Sergey in the spring of 1995, they connected instantly. Despite their differences, there was no denying the chemistry between them; the energy was palpable. It was during a new student orientation at Stanford, and Sergey was showing Larry and other prospective graduate students around the sun-laden California campus and its environs. Suddenly, the pair started to argue about random issues. It seemed an odd moment for two people who barely knew each other to be debating, sparks flying, but in fact each was playing a favorite game.*

In 1997, Stanford faculty, staff and students began to use the search engine created by the pair at the web address google.stanford.edu. Through word of mouth, their search engine quickly became popular. The size of the database and the number of users soon exceeded the capacities of their computers. But they were unable to afford new computers, so they looked for old computers no one wanted in warehouses and bought parts to try putting together devices. They once considered selling their patent.

Fortunately, they met an angel investor. Google also seized the timing of dot com bust by hiring many brilliant engineers from companies that got in trouble. For example, Alan Eustace (who would become my boss) and Jeff Dean (who invented the key back-end infrastructure at Google) both came from the now-defunct Digital Western Research Laboratory. Also, Google became such a legend that many creative engineers accepted pay cuts to join the company for its potential.

Google became more and more famous in Silicon Valley. Those who were able to get job interviews with Google would be labeled as "smart people." Computer engineers who never got a call from Google recruiters were embarrassed to admit it. That made Google jobs all the more desirable.

Microsoft employees were using Google products and talking about Google like everyone else. Some were skeptical, wondering how much money a search engine could make, saying "the kids in Disneyland" would eventually grow up and see the tough reality of the real world.

However, hundreds of Microsoft employees left for Google every year.

I had friends who became happier after joining Google. They looked more energetic than ever, and they said to me, "Kai-Fu, come to Google. It's fun!"

"Really?" I asked.

"Yes!" they replied with excitement. "I feel 20 years younger!"

I was impressed by how much they seemed to enjoy their work. Even more impressive was Google's so-called "free and transparent" culture. Employees chose what they wanted to do and formed teams based on their own interests. They all focused on how to make the company better and the products more useful. There were no secrets, no hidden agenda, no politics, no bureaucracy.

There was a well known story about Google's egalitarian working environment: a new Googler was unable to find an empty desk until he saw one in the CEO Eric Schmidt's office. He moved right in there. Schmidt was reluctant and made a suggestion in a timid voice, "Would you please ask around to see if there are other options?" The employee followed the suggestion but came back to say, "I asked everybody. They all said I should just stay here." Then the employee shared Schmidt's office for the next six months, during which Google was in the process of going public. Every time Schmidt received a phone call with confidential information, he had to find another place to talk. This continued until Google moved to a new compound. Schmidt purposefully picked an extremely small office so he could finally get privacy.

Such a story probably wouldn't happen anywhere else.

Most astonishing of all was Google's unique IPO process. Unlike other IPOs which favored larger investors, Google insisted on an auction-based IPO that treated small and large investors alike, providing everyone direct access to its shares. This offended numerous investment banks but won applause from the general public. Google put the user's satisfaction before profit. Most of its software and services were free. It would continue making products that were not necessarily profitable as long as they were useful. This was so rare in a profit-driven business world that it made me deeply admire the company.

But I didn't expect this admirable company to create a position that happened to match my goals. To an observer from the outside, Google appeared not all that interested in the China market, with no office in China and only a few engineers in the US working on its Chinese products.

One day in May 2005, as I was browsing news on sina.com, the most popular Chinese language website, a headline jumped into my sight, "Google Will Make it Big in China!" Then I learned from the news article that Google had purchased google.cn as the domain name of its future China branch, and was in the process of building a big "China plan".

Going back to China, joining a very cool company, doing groundbreaking work...Wasn't this job opportunity in front of me exactly what I was dreaming of and looking for? I believed in reaching out to grasp opportunities when seeing them, so I didn't wait. I reached out by looking up Google CEO Eric Schmidt's email address on line and writing to him about my interest in Google China. I was following my heart.

Google responded swiftly and efficiently. Alan Eustace, Google's senior vice president of Engineering and Research, reached me by phone that night. He sounded as warm as spring weather.

"Kai-Fu, I'm really surprised that you're interested in us," he said. "We're ecstatic about your email. Actually, we've been studying your background. We know you've worked in China and created a miracle for Microsoft Research Asia. You've worked in R & D for great companies. We've already had an internal discussion about you. You know what our senior vice president of Product Management, Jonathan Rosenberg, said about you? He said, 'We should circle him like wolves. He's a superstar!'"

## An Interview on the Golf Course

Even though I now knew Google wanted me, I wasn't sure if they could offer me an ideal position. Would it be a position in charge of the entire Google China? Would they empower me to make decisions? Would I have resources to develop Chinese products?

To get all my questions answered, I had the following conversation with Alan Eustace:

Me: I'm indeed interested in Google. But I'd like to know how big a team you are planning to build in China.

Alan: We're flexible about this. If you go, we're willing to build a team of more than 500 people.

Me: That sounds great. I am sure we can build a very strong team. Would you empower that team to build products suitable for China?

Alan: I'm sure that would be no problem. We highly value the China market and hope to do well there.

Me: Good. Having seen Microsoft flounder in China with poorly coordinated presence, I also want to make sure that you are prepared to build one Google China subsidiary, with me as the sole president responsible for the effort.

Alan: I am responsible for engineering and not other functions, but I understand your question. Why don't we fly you out for an interview? You can ask Eric [Schmidt] that question yourself.

Two days later, I received Google's invitation. But interestingly, the interview was not going to take place on the Google campus, but at Michael's Restaurant on the Shoreline Golf Course. The company chose a scenic and quiet location for the interview because there were hundreds of former Microsoft employees working for Google and they might recognize me. This was to protect my privacy.

I flew to California for the most intriguing interview of my life on May 27, 2005. It was a gorgeous sunny day. The sky seemed bluer and the grass greener than ever. On my way to the interview, I passed by a spectacular lake and greatly enjoyed the scenery.

The first interviewer I saw was a Chinese engineer named Niniane Wang.

The interview was surprisingly in-depth. She started with her favorite interview question, "How would you write a short program to tell if an image was a banana or an apple?" I outlined how shape, proportion, and color could be computed and tested quickly. She then asked me how I would improve Google search. I told her I would use user feedback. She probed further on how to measure feedback, and how to use these as signals for ranking. This was an area I had given much thought as a direct parallel from my specialty, speech recognition, where "speaker adaptation" would also use user feedback to improve recognition quality.  She finally asked about how an international company could build products that could compete locally.

Later I found out this is a unique Google interview method – the company always evaluates technical executives by having them interview with an engineer. Google engineers are geniuses who will not accept a technical leader who cannot speak their language or gain their respect.

Also later I found out that this engineer turned out to be a true genius who was featured in Google's recruiting ads as a poster child for her entering CalTech at age 15, and becoming a Microsoft games studio 3D expert before she was 20. After I joined Google, Niniane became a good friend, who helped our recruiting and internal training greatly.

During my interview with Niniane, CEO Eric Schmidt entered the room quietly. With a big smile, he said hi to me.

After Niniane left, Eric started chatting with me.

"Kai-Fu, this is not a formal interview," he said. "My mission today is to understand any concerns you may have, and to learn how you see international companies in China. Why do they always fail?"

"Eric, you know all the American Internet companies met their Waterloo in China. They operate poorly over there. I wonder if Google may fall into the same trap," I expressed my biggest concern.

Eric nodded and said, "We hope Chinese users find global messages through Google and get to know it's a good brand. We are planning long-term projects in China. We won't force you to do anything short-sighted."

"Our bottleneck is not that we can't get more users," he continued. "It's that we don't have a Chinese team to create the technology and interface Chinese people need. I believe you can recruit the best and brightest Chinese engineers for us to make the most needed products for Chinese users. How do you think we can succeed in China?"

"The most important thing is to give the local team liberty to do everything. Also, be patient to make a long-term investment. I've published an article about how to succeed in China called 'Making It in China.' I'll email it to you. It explains the requirements I think international companies must meet to develop their business in China. I hope you'll give it some thought. If you can really understand how difficult and special it is to succeed in China, I'll be happy to take on the challenge."

"OK, I'll read your email thoroughly," said Eric with a smile.

After I met with several vice presidents, Google co-founder Larry Page arrived in a casual outfit on his bike. He put the bike aside and walked in. When he looked around and didn't see Sergey Brin, he murmured, "I knew Sergey wouldn't be on time. He's always late!"

Larry combed his hair with his hand and sat down to ask me questions. His questions included the following:

"Do you know our hiring bar?  Given our hiring bar, how many good engineers can we hire a year?" "What do you think about Google's Chinese language products? How can we improve them?"

During my in-depth conversation with Larry, Sergey ran into the room with sweat dripping from his forehead. He was in a skin-tight purple cycling suit, carrying a skateboard. Obviously he came by skateboarding.

"They are two Peter Pans," I silently told myself. In the meantime I continued to answer Larry's questions. Sergey naturally joined in. The interview went on in a pleasant atmosphere.

All of a sudden, Sergey asked me, "Sorry, do you mind if I stretch?"

I wasn't sure what he was asking, so I asked him, "Sorry, Sergey, you mean you want to smoke?"

"No," he shook his head. "I asked if you'd mind if I stretched."

He wanted to stretch during the interview. I certainly didn't mind, just feeling very surprised.

Sergey sat down on the floor, stretching and asking me questions at the same time. This was really an unprecedented interview experience to me!

No wonder several people I knew seemed younger after joining the company founded by these two Peter Pans! I felt rejuvenated when looking at them and talking to them, too.

The two Google guys still dressed and exercised like college students despite their celebrity status. After the interview, they walked out with a hand on each other's shoulder like athletes at the end of a game. I overheard them saying, "People like Kai-Fu don't grow on trees."

That night Alan took me to an Italian restaurant for dinner. We exchanged opinions on many topics but carefully avoided technical issues of each other's company. I had learned by now that Googlers were all relaxed, friendly and pleasant but knew how to protect their business secrets carefully.

After a long day of interviewing, I felt a little tired on my way back to the hotel. But as soon as I entered the room, I saw a vase of flowers, and their fragrance immediately lifted my spirits. A card was attached to the flowers. It read, "Welcome to visit us in California! Best Wishes, Eric"

There was also a full basket of snacks and bottled wine in the middle of the room. The card attached to the basket was from Alan.

When I returned home in Seattle, an extra large package had arrived. It was from Google's Human Resources. I opened the box and saw all the Google toys, including a pen, a basketball, a USB flash disk, a T-shirt, a chair and the coolest of them all, a coin-operated miniature gum ball machine with Google logo.

I hid the big box of toys in the closet of my study. But before I tried to surprise my daughters with it, my youngest daughter Cynthia discovered it.

She came up to me, put herself on my lap, and asked in a still baby-like voice, "Daddy, are you going to work for Google?"

"Would you like Daddy to work for Google?" I asked.

"It'll be fantastic if you are really going to work for Google," she said excitedly. "I already liked Google when I was a little kid."

I couldn't help laughing. "You are still a little kid!" I said, looking at her still somewhat baby-like face.

"Daddy, do you remember there was a picture of doggie poop in the birthday card I made for you when I was three?" she asked.

"Of course I remember," I relied fondly, recalling that funny episode of our family life.

"I did it because you didn't let me get a dog but I really, really wanted a dog. I found a picture of dog poop on Google and put it in your birthday card. See, I started using Google when I was so little. I think Google is the coolest company in the world!"

Her cute expression made me laugh.

In the meantime my oldest daughter Jennifer came into the study and joined our conversation.

"Daddy, I think Microsoft is pretty good, too, with products like MS Word and Excel," said the 13-year-old who used the computer every day.

"Who wants those things?" retorted her little sister. "I think Google is the best!"

I smiled at their heated discussion. But at this time I wasn't sure whether I was really going to join Google. I didn't know what kind of offer Google would give me or whether the company would be able to enter China with an understanding of Chinese culture.

From my study, I emailed my article, "Making It in China," to Eric. The article explains how I believe international companies should operate in China. I hoped Google would take my suggestions to run Google China. But I wasn't sure how Google was going to respond. I told myself to wait and be patient.

## "Kai-Fu, We Are Being Sued!"

While waiting for Google's offer, I decided to use the sabbatical I had earned from Microsoft.

Microsoft gives every senior employee who has worked six years for the company a sabbatical, which will expire if the employee doesn't take the time off in the seventh year. It happened to be my seventh year with Microsoft in 2005, and my passion for the job was wearing off. I needed a break.

I applied for the sabbatical at the end of May 2005. My boss Eric Rudder approved it.

At an early June meeting, Rudder announced, "Kai-Fu will take six weeks off starting the 9th of this month. We'll have someone fill in temporarily."

Then he turned to me and asked half-jokingly, "Kai-Fu, you won't be like some people, gone after the sabbatical? You're coming back, right?"

Everyone was looking at me.

At that moment, I was considering leaving but hadn't made a decision. Also, I thought I would come back to ensure a smooth transition of duties even in case of resigning. Therefore, I intuitively responded in a low voice, "Yes."

Little did I know this simple answer would become the focus of a debate in court. A disaster was looming, but I was blissfully unaware of it.

I spent my sabbatical with my family. I visited my mother and sisters in Taiwan. My mother was overjoyed to see me. Every day I was there she made me my childhood favorite, mini dumplings with spicy Sichuan sauce. My sisters also expressed affections through food. They kept bringing me snacks from Shih-Lin, the most famous night market in Taipei.

One day in June, I went shopping with my family. Suddenly my cell phone vibrated in my pocket. It was a call from Google.

"Kai-Fu, I've got you an offer I believe you can't refuse! In terms of compensation, your salary and stock would exceed your Microsoft income, even if Google's stock doesn't go up. And you know it will!" said Alan with excitement in his mellow voice.

"What about my responsibilities?" I asked.

"We'll let you take charge of the entire Google China office, and give you the title 'President, Google China'. Each leader in the office will still have a functional boss, but you'll be the one leading the whole office. We want to give you maximum empowerment," Alan replied.

There was no way I could resist such an offer. I decided to join Google.

I ended my vacation early and returned to Seattle on July 2 to prepare for my resignation. On July 4 I invited a good friend of mine to my house and told him about my future plans. When I saw him off in front of my house, fireworks just started. Looking at the stunning splashes of colors in the dark sky, I wondered if the upcoming new chapter of my life would be as splendid as the fireworks.

I walked into Eric Rudder's office at 9 a.m. on July 5 and said, "Eric, I've been thinking about whether I should leave Microsoft. I've been with Microsoft for seven years. But now I'd like to work in China and Google is opening a China branch, so I'm going there with Google. Today I'm here to resign. I'll stay until completing the transition of my duties."

My boss first gave me a silent response. During those few seconds, the air seemed to be frozen in his office.

"Kai-Fu," he finally spoke up. "It sounds like you've made up your mind, but at least you should give Bill and Steve a chance to try keeping you. Don't talk about resigning now, OK? You must talk to Bill and Steve about this face to face. Bill is on vacation now. If I tell him Kai-Fu's gone when he comes back, he won't take it!"

"OK," I agreed. "But from today I'll stop reading Microsoft emails. I won't touch anything related to the operations of Microsoft, either."

"That's fine. We trust you," said Rudder.

I met with Steve Ballmer on July 8, 2005.

"Kai-Fu, if your ideal workplace is in China, you can pick a position in our China branches," he offered.

"Steve, I don't think Microsoft has a suitable position for me in China," I politely declined.

"Then what would you like to do in the headquarters? You can have your pick."

"Steve, I really thank you for the past seven years I've been with Microsoft. I've learned a lot," I tried to explain. "But I need to follow my heart."

Ballmer has a strong personality, which came out in this moment. He began to play a hard ball.

"Kai-Fu, if you are really going to Google, we have to take legal actions," he said firmly. "Don't think this is against you. We recognize your huge contribution to Microsoft. It is not you we are after. It is Google."

I was shocked, "But what I'll do for Google is a completely different type of work, Steve!"

"Just don't go," Ballmer said in a softer voice. "You can pick a position at Microsoft. Think about it. I'll arrange a new position and increase your compensation. Give me a few days, OK?"

I became silent. I realized leaving wouldn't be as simple as I had thought. I touched the most sensitive, vulnerable part of Microsoft's ego. Microsoft had been the dream company for software engineers, but Google was stealing its limelight. Over the past few years, Microsoft had lost several hundred people to Google. Microsoft and Ballmer could not accept that.

After thinking through what was going on, I said to Ballmer, "OK, Steve. Thank you for trying to keep me and thinking so highly of me. I'll wait to hear from you, before I make my decision."

I walked out of Ballmer's office with a heavy heart. Then I immediately went to see Rick Rashid, senior vice president of Microsoft Research, who had been my professor and my boss at Microsoft Research China.

My mentor expressed concerns after hearing my description of the conversation with Ballmer. "Don't leave," he said somewhat anxiously. "If you do, Steve may really take you to court!"

I thought Ballmer's mentioning of a lawsuit was an empty threat until getting this warning. Would that happen? I was only following my heart to change jobs. How could it bring me legal trouble?

I decided to call around for legal advice. To my surprise, all the large firms were unable to take on clients who might be sued by Microsoft. Microsoft had cleverly given business to all the major firms. Eventually, I found a small legal firm who didn't have Microsoft business. The lawyer said, "Kai-Fu, you signed a non-compete agreement with Microsoft, which prevents you from doing the same kind of work for another company within one year after leaving Microsoft. But as you said, there's no overlap between your Microsoft job and the work you'll do for Google China. Also, this law is enforceable in Washington but not enforceable in California. So the risks are not large. Besides, there are already more than 400 former Microsoft employees at Google. But Microsoft never sued anyone. So, Kai-Fu, you don't have to worry too much."

I felt relieved after listening to his analysis. I was determined to leave Microsoft. But I hoped Ballmer would gradually see my point. Perhaps this was just hard for Microsoft's top-level management to take for the moment. They would eventually accept it, and it would be just be a matter of time.

Ballmer called me on July 13, 2005. He said, "Kai-Fu, we've thought about it. We can create a new position for you and your title will be 'Chairman of Microsoft R&D China.' We'll also give you more stock options. So don't leave!"

This offer would fulfill my wish to go back to China and they increased my compensation even though I didn't ask for it. While I never disclosed Google's offer, the Microsoft offer matched it. These guys were experienced in giving counter-offers to Microsoftees leaving for Google! To me, however, the value of a job could not just lie in the monetary compensation. The Microsoft offer wouldn't give me enough space for new development or professional growth.

"Steve," I replied. "Thank you very much for the offer, but I don't think I will likely change my mind."

Ballmer still wouldn't take no for an answer.

Bill Gates returned from vacation on July 15, 2005. We met in the office the same day. He wasn't as hard-core as Ballmer, but he clearly expressed his disapproval of my resignation. He said, "Kai-Fu, Steve will take this to court. You know, Steve didn't sue those hundreds of engineers who left for Google because they didn't have your high seniority and people would tend to sympathize with the underdog. But you are a vice president. Steve thinks only suing you can stop Google from taking our people away!"

Oh no! With the good cop and the bad cop, Microsoft was doing everything to block my way to Google. But my principle told me not to give in. I heard a strong voice from my heart: I was doing nothing wrong!

After my talk with Gates, I flew to California again, but feeling depressed this time. I went straight from the airport to the office of Google's senior vice president and chief legal officer, David Drummond. Several lawyers were already there in his office.

David said, "Microsoft called us a few times, asking us not to hire you. They hope to settle this and have you stay at Microsoft. We're talking right now, and we estimate there's a chance of going to court."

"You know, you are influential in China, especially to young software engineers," he continued. "If Chinese people all follow you to Google, that'll be unthinkable to Microsoft."

I was silent for a moment, and then I said, "I really don't want to see my former employer in court."

"Don't worry, Kai-Fu," said David in an assuring tone. "First of all, you're most likely to win this case. We're standing on solid ground. Besides, Microsoft is based in Washington and Google in California. The State of California doesn't recognize the non-compete agreement."

With his assurance in mind, I returned to Seattle. I wrote my resignation letter in the Microsoft office on July 17, 2005. But it dramatically turned out later that I never submitted the letter. It still stays with me today.

In the letter, I wrote:

*I am writing to submit and explain my resignation. Thank you for spending so much time with me in the past two weeks, and for showing so much flexibility by accommodating my desire to move to China. It really made me feel special, and made my decision very difficult.*

*Ultimately, what tipped the decision was: (1) the opportunity to do a start-up -- I miss the flexibility and the camaraderie of a small team, and (2) the opportunity to learn -- by building not just R&D but a whole subsidiary, and driving the end-to-end presence in a country.*

*I hope we can have an amicable parting. You have my word that I will abide by the non-solicit and non-disclosure commitments I made when I joined. Also, as my book emerges as a best seller in China, I am using Microsoft to illustrate many positive angles of management, leadership, and corporate values.*

*The last seven years at Microsoft have been a very memorable learning experience for me. Thank you for giving me the opportunity to participate and learn here.*

*Kai-Fu*

After finishing the letter, I cleared my personal belongings from the office. I took a painting off the wall, my daughters' pictures from my desk top, and miscellaneous personal things from the drawers. I put all of them except for the painting in a cardboard box. With mixed feelings about leaving, I carried the box and the painting to the company garage and put them in the trunk of my car.

I didn't know I was being filmed in the garage. Later in a deposition, when I saw the video image of my lonesome-looking self walking into the garage with a cardboard box in one arm and a framed painting in the other, my heart ached. When the Microsoft lawyer asked me if the box contained any Microsoft confidential documents, I tried very hard to hold back my anger and my tears.

I planned to submit the resignation letter on July 18, 2005 after a meeting with Microsoft Fellows, a group of Chinese students who had been selected from Microsoft Research Asia to visit the Seattle headquarters. I had pleasant and smooth conversations with the students in the conference room. I meant to do my last job at Microsoft well. But my cell phone vibrated in the middle of the conference. I quietly took a look: area code 650, from Silicon Valley, so I excused myself from the conference room to take the call.

"Kai-Fu, we're being sued!" That was the first thing I heard from the call.

Microsoft had filed a lawsuit against me and Google for violating the non-compete agreement.

The news astonished me. I hadn't even submitted my resignation letter! How could they sue me before that?

Microsoft, the company I had served for seven years, the company I had striven for and fought for, would choose to repay me this way upon my departure! What would this do to my future and my dreams?

What would this do to my reputation, which I cherished? Would this hurt my family or my children? Would this affect my career prospect? I felt completely lost...

This was probably the most dramatic moment of my life, unfortunately an utterly tragic one. Now looking back, I can still feel the bone chilling despair I felt then.

"What if I don't submit the resignation letter? Will that make Microsoft a big joke?" This thought flashed through my mind. But then I rationally told myself not to do anything out of spite. The job change was a serious decision I had made for my future.

I suppressed all my hurt feelings and returned to the conference with Microsoft Fellows. We continued to talk about the company that was going to take me to court.

At the end of the conference, the students cheerfully asked me to be photographed with them. I tried extremely hard to smile to the camera. The students surrounding me had no idea how much pain I was going through during that Kodak moment.

*With Microsoft fellows at Microsoft headquarters*
*July 18, 2005*

That afternoon I printed out my resignation letter and the plan of job transition. I took heavy steps toward Eric Rudder's office with the documents. Then I saw a few people in business suits standing outside of that office. I immediately realized they were lawyers.

Rudder was waiting for me in his office. When I walked in, he said, "Kai-Fu, please have a seat."

I sat down silently. He looked a little embarrassed. It seemed difficult for him to find words, but he finally started to talk. He said, "Kai-Fu, I think you'll eventually join Google. The lawsuit thing will only hold you back for a while. It'll blow over sooner or later."

"Eric, I know you are not involved with this," I said sincerely. "Since this is already happening, I'll deal with it. Today I'm here to resign and to talk about the transition of my job. I'm willing to spend time on the transition of duties. It's my responsibility."

"Oh, don't worry about it," he responded kindly. "With this thing now, Kai-Fu, you need to focus on your lawsuit. Don't overburden yourself with the job transition."

"Thanks, Eric. I've written a plan for the job transition. If you don't need me involved with it, you can take this for reference." I handed him the plan.

As soon as I walked out of Rudder's office, one of the lawyers waiting outside approached me and gave me a manila envelope.

"Mr. Kai-Fu Lee," said the lawyer. "Because of your new position with Google, Microsoft is suing you for violating the non-compete agreement. Please sign your name here to accept all the lawsuit documents. You are hereby served."

I followed his instructions. Then I saw the law firm's name on the manila envelope---Preston, Gates and Ellis. The name Gates rang a bell. Yes, one of the law firm's partners is Bill Gates' father.

# Appearing on Every Magazine Cover

Microsoft sued me for violating the non-compete agreement, and Google for intentionally assisting me in the alleged violation of the contract.

It must have taken weeks to prepare documents for such a lawsuit. Microsoft must have started all the legal paperwork while they stalled my departure by asking me to talk to everyone and wait for their counter-offer. They needed to launch the lawsuit while I was still a Washington resident. They were afraid that once I moved to California, I would become a California resident and the jurisdiction would change to California, where non-compete lawsuits were not allowed.

Years of experience with anti-trust lawsuits had made Microsoft a legal powerhouse. Microsoft's Legal Department was bigger than all of Microsoft Research worldwide. The vice presidents in Legal outnumbered the vice presidents in Research.

How was I, a scientist with expertise in technology but little knowledge of law, going to fight against such an army of experienced lawyers? I just lost round-one – while I naively waited for my counter-offer, Microsoft bought the time needed to sue me before I left.

I felt hopeless and helpless. Why me? What would this do to my reputation, family and future? In the sunshine of Seattle's July, I was only feeling cold.

What worried me most was the impact of the lawsuit on my family, to whom I was always a pious son, caring husband and loving father. My successful career had made them proud. But would the lawsuit change everything for them? In Chinese culture, lawsuits are frowned upon and the accused often stigmatized. How could my elderly mother take the news of me being an accused? How much would my wife suffer from my suffering? Would my children become targets of school gossip because of me?

"Can we settle out of court with Microsoft?" I asked Google's lawyers. I thought it would be for the best to avoid fighting my former employer in court. It just seemed too cold-blooded for me to become enemies with people I had worked with for seven years.

David Drummond, the 6'3" chief legal officer with a demeanor resembling Martin Luther King's, provided his expert opinion after listening to me. He said, "Kai-Fu, it would be ideal to settle out of court, but Microsoft is using this lawsuit to have other employees who may want to leave for Google see you as an example. Obviously they are not simply trying to stop you. They want to prevent other employees from pursuing the same dream. So while we try to settle, I believe the chance is very slim. We need to get ready for a hard fight."

"What should we do now?" I asked.

"We'll announce your coming to Google in a press release. Let's spread the news," he said in a firm tone, his eyes sparkling, which boosted my confidence in winning the lawsuit.

On July 19, 2005, the announcement instantly became headline news and cover stories in the United States and China.

Bing-Lin Gu, president of China's Tsing-Hua University, also appeared in the news in China with his strong support for the Google announcement. He said, "Dr. Kai-Fu Lee possesses the perfect combination of technical brilliance, leadership excellence, and business savvy, and he cares deeply about the students and education in China. Dr. Lee is the ideal candidate to do great things for Google and for China."

Only minutes later, the Microsoft version of the story spread through the Internet as quickly as virus. Headlines such as "Microsoft Accuses Lee of Violating Non-compete Agreement" and "Microsoft Takes Lee to Court in Washington" appear everywhere. Those articles were questioning my loyalty.

Microsoft told the media that I "resigned out of the blue." But what happened was Microsoft sued me out of the blue. The fact that they filed the lawsuit before I resigned made me wonder if their intention was to buy time for preparing the lawsuit when they made offers to keep me. A friend of mine working for Microsoft later confirmed my speculations.

"Microsoft was indeed getting ready to sue you when they asked you to stay," he said. "They made me and some other people friendly to you leave after a meeting that discussed how to keep you, and then they would have another meeting to brainstorm about how to sue you."

However, this episode was not in the news. People only knew two high-tech giant companies were fighting over an executive. The high profile lawsuit turned my private decision of a job change into a public affair.

I was shocked to see headlines such as "Microsoft Says Lee Planned to Jump Ship Long Ago," "Microsoft and Google Fighting for China Market," and "How Much Did Google Offer Lee?"

Tom Burt on the Microsoft legal team told the New York Times that I held trade secrets pertaining Microsoft's search technology and China business strategy, which would prevent me from working for Google China based on my non-compete agreement with Microsoft.

"It's a very egregious violation of his non-compete agreement," said Burt of my new position with Google.

In the meantime, Google fought back, stating that the company had reviewed the Microsoft claims and found them "completely without merit." A Google spokesperson said, "We're thrilled to have Dr. Lee on board at Google and we will defend vigorously against these merit-less claims."

As the fight went on, more and more fingers were pointed at me.

Some speculated whether I had been involved with Microsoft's core business secrets, especially in search. Some did not realize the limited scope of the non-compete, and criticized me for breaching the agreement. These completely false accusations bothered me as I worried about their possible impact.

In the meantime, Microsoft provided many documents to the court and to reporters. One document exposed my Google offer as well as my Social Security number. Another alleged that I received a million dollars as a payment for signing the non-compete. Critics were quick to write that I should not have left after receiving so much money.

"That's absolutely nonsense!" I expressed my anger to Google's lawyers. "The one million dollars I earned in 2004 all came from my salary and stocks. That was my annual income. It had nothing to do with the non-compete agreement."

From that moment I realized I was already on trial even before going to court. My integrity was being questioned and judged. All the rumors about me on the Internet were harsher than any cross-examination.

A college student wrote me a public letter in Chinese, titled, "To Kai-Fu Lee---Let's Start with a Discussion of Honesty and Integrity." It mocked the public letters I had written to college students in China and portrayed me as a traitor to Microsoft. This letter became widely circulated on the Internet, at the expense of my reputation.

The letter and other articles attacking me were all published without fact checking. I wanted to defend myself, but my lawyers told me that I must stay completely silent as the accused, so I didn't make any statement.

Many of those who accused me of violating the non-compete agreement didn't even know what the contract really entailed. By law, an agreement like this is for companies to prevent former employees from doing *the same kind of work* for a competitor. Microsoft's non-compete agreement will restrict former employees for a year after they leave.

Not all of the states recognize the non-compete agreement. As David Drummond once explained to me, Microsoft's home state Washington does, but California, where Google is based, doesn't. Furthermore, even in Washington there was a precedent, *Perry v. Moran*, which made clear that employers cannot use the non-compete to broadly limit their former employees.

In my case, what I did for Microsoft was all about speech recognition, natural speech language software, and a help system. The non-compete agreement would only keep me from working in those areas for Google.

As for the well known competition between the two companies in search, I was absolutely not involved. When I worked for MSN, Microsoft didn't even have a search engine.

The closest I had been with Microsoft search was the help system my team developed. This help system looked up help topics for users. When the user entered a request such as font enlarging or photo downsizing, the system would match the request with one of the functions available. This kind of search was within a software system, completely different from Google's Internet search technology.

However, these technical concepts were difficult for non-tech people to understand. Microsoft's PR machinery took full advantage of that.

Fabricated news spread everywhere. Before one of my frequent Seattle-San Francisco flights, I walked into an airport bookstore and saw pictures of myself on almost all the business magazine covers. That made me walk out immediately. But after I sat down on the plane, I picked up an airline magazine and saw reports on my case again!

Pressured by Microsoft's exaggerated claims, the King County Superior Court in Seattle announced a temporary ruling before sorting through the overwhelming amount of technical documents. The ruling prohibited me from taking my new position with Google until the official court hearing on Sept. 13.

Facing the discouraging temporary ruling, the frenzied media, and the relentless Microsoft, I felt I was knocked out again in the second round of the fight. The game was not over, but how many boxers could win after two KOs?.

I was exhausted, but unable to go to sleep. The lawsuit was a nightmare waking me up every night. Looking into the darkness of the night, I often wondered: when would I ever see the light at the end of the tunnel?

## The Most Difficult 60 Days

After the temporary ruling, more and more groundless accusations came from Microsoft, which hurt me more and more deeply. How could I be accused of so many crimes I had NEVER committed? I kept asking myself: how should I control my emotions under such tremendous pressure? How was I going to deal with the most difficult hardship of my life? How was I going to make people see the truth when all they received was false information?

Sometimes I wondered how Microsoft came up with so many imaginary stories. It wasn't until one and a half year later that a former public relations manager of Microsoft China told me what happened: to ensure my reputational damage in China regardless of the lawsuit outcome, Microsoft China formed a PR team to work on lobbying analysts, writers, and bloggers to write against me, and lobbying the Chinese press to tilt its coverage against me. One team member actually had worked with me in China and knew my personality. She couldn't stand spending day after day spreading untrue rumors about me. One day she ran out of a meeting room with tears in her eyes.

I'm not sure what happened in the Microsoft headquarters, but perhaps something similar.

One of the major fabrications was about my involvement with leadership at Microsoft search. They claimed I was a search expert, managed search technologies, and knew Microsoft's search secrets.

Microsoft told the judge that if I was allowed to work at Google, there would be a risk that all of its confidential search technologies would walk into Google. Microsoft also claimed that I was such a search expert that Bill Gates had requested four to five meetings with me to discuss search strategies. All of these were fabricated.

Microsoft told the media that I said yes to my former boss when he asked me if I would come back after my sabbatical. They said I lied because I signed paperwork that promised to come back after my sabbatical. But later it was discovered that on my signed form, there was no such promise to return. The lie came from Microsoft, not me.

Microsoft also questioned the purpose of a talk I gave in June to a delegation of visiting Chinese university presidents in Seattle. The university presidents sought me out because of my influence with Chinese students, yet Microsoft claimed that I was using the company's name to befriend Chinese university presidents. In fact during that event I talked to China's ambassador Zhou Wenzhong, the other keynote speaker, about arranging for China's President Hu Jintao to visit Bill Gates' house -- so I was actually using my influence to help Microsoft, rather than the other way around. I didn't expect Microsoft to give me credit for that after my departure. But how could they use it as a reason for attacking me?

In addition, they said I had recruited talents for Microsoft Research China and that would prohibit me from doing recruiting work with Google. They said I knew "recruiting secrets" from Microsoft. What secrets were these?  When my lawyer asked a Microsoft senior vice president later in court, he was only able to say, "Hire people---some experienced, some less experienced. Hire tens, hundreds, or thousands of people. Hire them from universities, hire them from within Microsoft, hire them based on referrals."

Could those common ways of recruiting count as business secrets?

I could have laughed at such ridiculous accusations. But I was too much hurt to laugh. Every morning I turned on my computer, and the first thing I saw was negative press about me. I lost my appetite and 15 pounds in those two months.

I didn't dare to tell my mother about the lawsuit. But I knew it was impossible to keep her from knowing. Taiwanese media went crazy about this case, too. It would only take turning on the TV set for her to find out.

One day my mother called me from Taiwan. She didn't mention a word about the lawsuit. She simply said, "Son, mom trusts you. Don't skip meals, OK? Take care of yourself!"

Tears rushed into my eyes, though I was raised to be a strong man never supposed to cry...

My supportive family gave me strength to carry on through those toughest days. In order for me to concentrate on the lawsuit, my wife didn't let me do any of the packing when we moved from Seattle to Silicon Valley. At first she single-handedly worked on it. But later she realized there were too many things in our Seattle home for her to pack up before our moving date. Then two of my sisters flew all the way from the East Coast and Taiwan respectively to help her. Thanks to their assistance, our packing was done in time.

Right before we moved, some of my Microsoft colleagues held a going-away party for me. They took turns expressing their appreciation of what I had done for them and gave me their best wishes. It was a heart-warming afternoon, at the end of which I was surprised to overhear them tell one another not to mention this party at work. Then I learned that Microsoft had made a rule forbidding all employees to keep in touch with me. These friends were risking their jobs to say good-bye to me!

Later I heard that one of them did get in trouble because of the going-away party. This friend was called upon by my lawyers as a witness, so Microsoft had a lawyer talk to him first. The Microsoft lawyer asked when he last saw me. He was a honest person who could not tell a lie, so he said, "A few days after Kai-Fu's departure."

"You've seen him after his departure?" The Microsoft lawyer continued to probe. "Where did you see him?"

"At his going-away party," my friend gave the shortest answer possible.

"Going-away party?" The Microsoft lawyer was surprised. "Who held a going-away party for him? Who else went?"

Then my friend refused to reply, and that made his life difficult at Microsoft. He soon decided to take an early retirement.

The abrupt ending of his career made me feel terrible. In the meantime, I felt guilty about no longer sharing household duties with my wife. After we moved, my wife did all the unpacking and all the housework by herself. I was afraid she was being overburdened. But whenever I asked her if she was tired, she always smiled and said she was doing very well.

My daughters didn't complain one word about what my lawsuit was putting them through, either. They went to new schools in Silicon Valley and continued to get good grades. They were as sweet to me as ever.

One day I saw a lot of lawsuit reports in the newspaper at home and assumed the 13 year-old Jennifer had read them. I couldn't help worrying about how she might feel and what rumors she might have heard in school, so I started a conversation with her alone. I wanted to explain everything to her but found it difficult. What words would be the most appropriate for communication with my introverted teenage daughter?

To my surprise, Jennifer said, "Daddy, you don't need to say anything. I understand. I've always trusted you, and I always will. You don't need to explain."

The maturity beyond her age she displayed stunned me. Her simple but significant words empowered me. My daughter was being so strong. I ought to be even stronger!

I told myself not to let rumors torture me any more. I reminded myself of my motto, "Have the courage to change what's changeable, have the magnanimity to accept what's not, and have the wisdom to tell the difference."

I decided not to waste one more second on the unchangeable rumors. Instead I focused every second of my waking time on the still changeable lawsuit. I still had a chance to win!

I stopped reading anything about the lawsuit and began to work hard with my lawyers. Google formed a dream team of lawyers for me. Each one of them was the cream of the crop in California. I met with the seven of them every day to collect evidence and review documents.

The dream team included three Google lawyers headed by David Drummond. The second one was Google's deputy general counsel Nicole Wong, a Chinese American, who is always calm and logical. The third was a bright young Korean American attorney, Michael Kwun, who was in charge of all the lawsuit's details.

Three other lawyers were top lawyers from the famous legal firm Keker & Van Ness. These three were headed by John Keker, who had handled numerous high profile lawsuit including former President Reagon's Iran Contra case and the famous investment banker Frank Quattrone's lawsuit. From John's team was a technology law expert of Indian descent, Ragesh Tangri, who knows enough about computers to work as a manager for a high tech company. There was also Susan Harriman, who is soft spoken but can turn into another person during cross examination. When she role-played cross examining with me, her relentless questioning made me feel as intimidated as facing a real Microsoft lawyer!

Finally, Brad Keller was my personal lawyer. Brad has a mellow and likable charisma that many could mistake him for a gentleman from a southern plantation.

The dream team boosted my confidence in the lawsuit. We worked together for six weeks, during which our first task was "discovery." We asked Microsoft for the emails I had written and received at work, as well as other Microsoft-internal documents, to be used as evidence.

Microsoft was legally obligated to provide those emails, so they did. But we received a lot more than we requested. They also gave us my former colleagues' messages related to my Microsoft emails. The total exceeded 300,000!

The 300,000 emails were in PDF files on 20 CDs. These files had no text but only pictures of the emails, as though they were photographs. That meant we couldn't use key words to search passages.

"Are we going to read these emails one by one?" My lawyers expressed their overwhelmed feelings.

Obviously Microsoft wanted to make it difficult for us. However, they forgot that I was a speech recognition expert. I calmly told my lawyers, "Don't worry. I'll solve the problem."

I helped my lawyers find an OCR (optical character recognition) program which converted the pictures into text. Then, we loaded all the documents into Google Desktop Search.

Unfortunately, the OCR tended to misread words once in a while. As a result, year 2004 often became year 7004, year 2005 was mixed up with 2006, and "Ballmer" usually turned into "Balder". The Word documents that came out of the scanning process were full of such mistakes.

The mistakes slowed us down at first. But I soon came up with a simple solution – since the confusions were consistent and not frequent, to overcome the confusion between 2005 and 2006, we could simply search for both numbers when looking for my 2005 emails. This way we wouldn't miss any important evidence.

While reading our search results, I was shocked to see an email from a former colleague at Microsoft Research China to a headquarters vice president. The email essentially said, "Congratulations! Your plan was brilliant. Chinese people value loyalty, and our PR team has successfully stigmatized Kai-Fu as a disloyal person."

I also found myself betrayed by a manager I had mentored at Microsoft Research China for years. He said in an email to the headquarters, "We hope to get more resources from you in order for us to beat the founder of Microsoft Research Asia in China. Imagine what a big win that would be!"

I thought these two people were my friends! I would never expect them to risk their jobs by standing up for me against their employer. But how could they act happily to help Microsoft attack me despite all I had done for them?

My heart ached. But there was no time for sadness. I had to put my emotions aside! I took a deep breath and then continued to look for evidence with my lawyers.

We found the evidence we needed. One email written by an employee of MSN Search said, "Though Kai-Fu Lee had nothing to do with our projects, we have to say he worked with us for the sake of the lawsuit. We'll frame him that way."

We also went through Gates' calendar as well as mine and proved the "four to five" search meetings claimed by Microsoft were fabricated.

We found these, and many other valuable pieces of evidence. My lawyers were very happy. One of them said, "Hey, Kai-Fu, if you ever want to change jobs again, you can work for us! You are as good as two people."

"Really?" I asked, making an expression of being flattered. "Am I really as good as two lawyers?"

He shrugged, "I meant two IT people."

In addition to collecting evidence, my lawyers suggested trying to get some fair news stories on the case so the judge wouldn't find all the articles pro-Microsoft when reading the newspaper in the morning. Particularly, it would be valuable to show that Google wanted to hire me for my inherent abilities, rather than what I learned from or did at Microsoft. This was another challenge. Since I wasn't supposed to do interviews, I needed to find someone objective and committed to not quoting me.

I recalled a young reporter I had met in China, Kristi Heim, who had worked in China and once followed me to a Chinese university to cover my speech there. With her excellent command of the Chinese language, she understood what I had done for Chinese college students. More significantly, she happened to work at the Seattle Times, the newspaper the judge probably would read every morning.

Would she write a fair report from an objective point of view? When I reached her by phone, she said, "Kai-Fu, I know you are a decent person. But as a news reporter, I have to investigate the case and let both sides voice their opinions."

In August 2005, an article titled, "Microsoft and Google Feud over Top Exec" appeared in the Seattle Times. The article describes me as "a kind of spiritual leader to aspiring Chinese technologists."

It quotes a Beijing University graduate named Gao Jian, "Kai-Fu Lee is a kind of idol among universities. Students in universities just want to choose a name and go where he is. Bill Gates and Kai-Fu Lee are both a kind of hero in the business, but Kai-Fu Lee has more contact with the students. And he is a Chinese. I think this is an important reason."

The article also analyzes the lawsuit and states, "But the real struggle may have less to do with Lee's technical expertise and more to do with his ability to influence a generation of young technologists, especially in China."

I hoped the judge would read this article and see the lawsuit in a different light.

Following the Seattle Times article, other media began to tell the Google side of the story. Google repeatedly announced that the company didn't need any of Microsoft's technologies and thus hired me not for technical know-how, but for my management abilities and my influence in China. Google also pointed to my record of never leaking business secrets in past job changes to prove my honesty and integrity.

While some accurate news and balanced reports made me feel better, my nerves were still all wound up, unable to relax. I felt as if skating on thin ice. Any minute I could fall through the ice and drown in deadly cold water. Every step was a risk to take.

## Becoming Enemies with Old Friends

On Aug. 10, 2005, my personal lawyer called me and asked, "Kai-Fu, how many computers do you have at home?"

"Two," I answered. "Mine and my daughter's. What's up?"

"Does your computer keep any Microsoft document?" he asked.

"Of course not. I didn't keep anything from Microsoft."

"That's good," he sounded relieved. "Someone will go to your house to take your computer. Microsoft has asked a third-party agent to look through your computer files. They want to see if there are any Microsoft business secrets on your hard drive."

"I don't have any Microsoft business secrets!" I shouted. "But how am I going to work without my computer?"

Two hours later, my computer was taken away. That caused a lot of inconvenience. Even though I bought a laptop later, I lost a lot of personal data saved on the confiscated computer, such as my tax returns, personal emails, music and photos.

A month later, the third party agent submitted a report, which stated, "No Microsoft documents were found on Kai-Fu Lee's personal computer."

The computer was returned to me with a broken hard drive, which inevitably upset me. However, I understood it was a necessary evil. Any kind of evidence anyone could think of had to be collected at this stage in preparation for the depositions.

A crucial part of any lawsuit, depositions give both parties a chance to publicly express themselves, question the other side and discover new evidence before the official hearing. In my case, depositions meant even more because I needed the truth to come out in public to dismiss all the haunting rumors.

On Aug. 26, 2005, I went to the Microsoft depositions, during which Microsoft executives provided their evidence and were questioned by my lawyers. After I left Microsoft, that was the first time I saw my former bosses and colleagues of the company again. I saw Bill Gates, Steve Ballmer, Eric Rudder and Microsoft's chief technical officer, Craig Mundie.

Gates never looked me in the eyes that day. I expressed my feelings about seeing him again in my journal:

*When Bill Gates walked into the room, he did not look at me. Was it because he sees me as enemy and won't talk to me? Was it because the legal team gave him coaching that this is the time to make me feel bad about my "betrayal"? So that it would negatively affect me in my deposition and at the preliminary injunction?*

*Here is one of his seven most trusted advisers, someone who once confided in him that: "Bill I would never lie to you. I want to tell you what I can do and what I cannot do." Someone who once rescued him from a disastrous meeting, and went to his suite in the hotel to tell him that he needn't worry any more – someone to whom he showed his most sincere and innocent smile of appreciation. Someone he really trusted. Now what has become of this person? I thought of Steve Ballmer's famous quote: "You are either with us, or you are against us"? (The famous words Steve Ballmer used with a customer who adopted Netscape Navigator)*

It also saddened me that day to see Mundie, who had taken his wife with him on a business trip to China during my time with Microsoft Research Asia. He asked me, "Can your wife help my wife look for garbage?" I was surprised. Then he told me about his wife's hobby of collecting second-hand things. My wife took her to the flea markets in Beijing and made her very happy. In the meantime I accompanied him to see Chinese government officials. He criticized a high tech policy of China in front of a Chinese minister, who knew little English. I was afraid he would offend the minister because the government's authority was hardly ever challenged in China, so I only translated about 20% of his words. After the meeting, he asked why the Chinese translation was so short. I told him, "Brevity is the magic of the Chinese language."

I described my deposition encounter with Mundie in my journal as well:

Craig Mundie walked in from his vacation. He had just returned from a long trip on his boat. He had a friendly smile when he saw me, and was eager to tell me about his trip. We talked about his boat computers, where he went, how he decompressed.

Then the reporter turned on the machine and Craig was a different man.

*In the morning, it was "Kai-Fu knew everything." And in the afternoon it was "Everything is a trade secret."*

I knew Ballmer would be the harshest among them all. But I was still shocked to hear his deposition. I described how I felt about it in my journal:

*He started by saying, "To our Chinese employees, Kai-Fu is like a Godfather."*

*My attorney Ragesh asked "Mr. Ballmer, is Godfather an official Microsoft title?"*

*I was unable to laugh at that humorous question, because Ballmer continued to say that*
*I had "broad and unilateral" responsibility for the company's China strategy.*

*What a huge lie that was! Responsibility, perhaps he offered me some before I left, but I didn't take that job! Broad? No way. Unilateral? Impossible. "Godfather !" Was this about the respect people had for me? Or the work I voluntarily did for Gates and Ballmer? Was my generosity turning into a monster to bite me?*

My lawyer Ragesh handed Ballmer a presentation, which was based on my article, "Making it in China," and used for my speeches at various business schools in the US. We intended to show that this presentation was all public information. But to our surprise, Ballmer said this document was confidential to Microsoft. He then dissected the presentation and stated how each section could be based on Microsoft's data. He made a special point that work built on data Microsoft purchased should be owned by Microsoft.

My lawyer asked, "How do you decide if something is confidential?"

Ballmer said, "Confidential documents are clearly marked 'Microsoft Confidential.' If it's not marked, it needs to be considered confidential unless approved otherwise."

"Who is authorized to approve?" asked my lawyer.

Ballmer said just him and the senior VPs. This was a rule previously unheard of, and I knew for sure a vice president was authorized to approve, but he seemed to be making up rules that removed my discretion.

Why did he go through such lengths to go on record to talk about Microsoft-owned data? And marking "Microsoft Confidential"? And saying I had no discretion? Could it be that their lawyers prepared something to trap me later?

After listening to the three depositions, my lawyer expressed concerns during recess.

"Obviously they've been trained by lawyers," he said. "The lawyers prepared seamless answers for them so they wouldn't leak anything that might be in your favor."

"What about their exaggerated and false statements?" I asked. "Can we use those to question their integrity?"

"We will," my lawyer said. "But this case is about your integrity rather than theirs."

I was losing hope. I recalled the last few words of the pledge of allegiance, "with liberty and justice for all." Where was my justice?

I didn't expect the situation to soon turn around with MSN Search vice president Christopher Payne's deposition.

In the Q&A with our lawyer, Payne essentially said, "Microsoft didn't really do search when Kai-Fu Lee was with MSN. Microsoft search was my idea after I joined MSN in 2002. We never invited him to any of our department meetings or our meetings with Gates. We didn't put his name in our letter of appreciation during the product launch because we didn't need to thank him for anything. I'm in charge of Microsoft search and Lee had nothing to do with it."

When he stated the facts with a sense of pride, the Microsoft lawyer sitting beside him suddenly turned pale.

Neither my lawyers nor I could believe he was actually telling the truth. His own desire for credit clouded the instructions he had been given – to incriminate Kai-Fu Lee as a search expert and manager. This was the most favorable evidence to us!

## My Deposition

Within the same week, I had my deposition, where the Microsoft lawyer started the session trying to intimidate me.

Lawyer: Dr. Lee, do you believe honesty and integrity are two of the most important virtues?

Me: Yes.

Lawyer: Would you please read the English version of the passage about honesty and integrity in your first public letter to Chinese college students?

Me: "I once interviewed someone who was talented in technology and engineering. But he told me during the interview that he would bring a product he had created when working for his current company if I hired him. Then he seemed to feel he might have said something inappropriate. He explained that he had made the product in his own time after work and his boss didn't know about it. But after this talk, to me, despite his abilities and work performance, I would never hire him. The reason is that he lacks the most basic principle and work ethic---honesty and integrity. If I hire such a person, who can guarantee he won't call what he does here something he has created in his own time and present it to another company as a gift? This tells us that a person without integrity cannot become a real success."

Lawyer: How do you feel after reading this?

Me: This is my principle, which I won't compromise. To me, honesty and integrity are more important than my life. This is the value I always keep every time I change jobs.

Lawyer: When you applied for your Google position, did you present a gift?

Me: Of course not. You can see in my emails that Google asked me not to talk about business secrets of Microsoft, and I told Google I would only do projects I never did for Microsoft. You can see, from Apple to SGI, from SGI to Microsoft, I've never had any problems with this through job changes. I've been extremely cautious.

Lawyer: Dr. Lee, did you know Google CEO Eric Schmidt before?

Me: Yes, I first met him more than a decade ago. But we didn't keep in touch.

Lawyer: If you were not in touch with him, how did you approach him?

Me: I sent him an e-mail.

Lawyer: You were not in touch with him. How did you know his email address?

Me: I Googled him.

Lawyer: Did you write an email to recommend a former Microsoft employee Alan Guo to Google?

Me: No. I just let him list me as a reference.

Lawyer: Wasn't Alan Guo a Microsoft employee?

Me: When I recommended him, he had left Microsoft for two years. He was about to complete his MBA at Stanford.

Lawyer: Then you were still helping Microsoft's competitor.

Me: No. I was Alan Guo's mentor for five years. That was why he listed me as a reference. I would be his reference for his interview with any company.

Lawyer: Have you recommended him to Microsoft?

Me: Yes. I did in 2001. Then he worked for Microsoft for two years.

Lawyer: Did you recommend him to Microsoft again when he completed his MBA at Stanford in 2005?

Me: Yes. He wanted to go back to China, so I recommended him to Microsoft China. But it didn't work out. I have emails that I can submit as evidence. Would you like to read them?

This type of tricky Q&A went on for seven full hours. When it was finally over, my coach Susan Harriman looked relieved. "Impeccable!" she exclaimed. "Kai-Fu, I've never seen someone like you in cross exam. All your answers were thoughtful, truthful and logical. You didn't fall into a single trap. We're all proud of you!"

I was proud of my lawyers, too. As I walked out of the court room, I felt grateful, not only to the dream team of lawyers, but also to my character witness, Xu Xiaoping, dean of the New Oriental Education and Technology Group's Research Center in China.

I once thought I could easily get a friend to be my character witness. But it turned out that the employers of my friends held them back for business relationships with Microsoft. When I was in despair, Mr. Xu (who keeps the Chinese way of having family name precede the given name in the English translation of his full name) came to my rescue.

He detailed my positive influence on China's college students in his written statement:

From my communication with students in the New Oriental School, I have learned that Dr. Lee has an enormous impact on many students in terms of their career planning. Dr. Lee's letters and articles help them establish their value systems. I personally often quote Dr. Lee when I give speeches. Dr. Lee's speeches are particularly popular with students from high school to Ph. D. level, between ages 18 and 40. The website of our school often publishes his articles, which enlighten not only our students but also others interested in studying abroad. Numerous students have told me they believe Dr. Lee is a venerable, trustworthy person from whom they can learn a great deal...

Mr. Xu even claimed he was willing to take legal responsibility for this testimonial. His deep trust in me kept me from losing faith in human nature.

In the stage of depositions, a friend working for Microsoft also stood up for me when he was called upon as a witness from my former employer. He attested to my integrity and said he trusted me in his deposition, regardless of all the instructions Microsoft lawyers had given him. His extraordinary courage deeply touched me.

While some old friends turned into enemies, I learned to hold true friendship even more dearly.

## Desperately Seeking Evidence

After the depositions, my lawyers analyzed them and told me the biggest concern was Microsoft's attempt to use my article, "Making It in China," as evidence that I gave Microsoft data to Google.

"Making It in China" was written during my employment with Microsoft, but only the version I submitted to Microsoft's top-level management contained my suggestions for the company's operations in China. I deleted all the Microsoft-related information before publishing it, and the public version was what I sent Google CEO Eric Schmidt.

"Do you have a witness to prove this version is public?" My lawyers asked.

I started a search in my memory. The article was distributed to many students during my speeches. But I didn't know any of those students. I could only look for people who had printed and duplicated the paper for me---Oh, yes, Mary Hoisington! My former secretary at Microsoft, who had retired before I left.

Mary had been my mentor Rick Rashid's secretary before working for me. When she followed Rick on one of his business trips to China, I was working there. I took them to a restaurant that served the cuisine of a Chinese minority ethnic group for the sake of novelty. The meal included some exotic food such as insects and snake. Mary seemed to enjoy everything a lot, so I thought I had made the right choice of restaurant. Only later did I find out that she was actually repulsed by the food, but she had learned from an article that in China guests would always praise the food as a way of showing appreciation to the host, so she decided to be thoughtful.

Given her thoughtfulness and kindness, I believed she would be willing to help me.

Indeed, after listening to my request on the phone, Mary said, "Yes, I remember copying this article. Also, Kai-Fu. I was sad to see all the negative press about you. I didn't call you only because I thought you would be too busy with the lawsuit to chat. I want to help you!"

After agreeing to be my witness, Mary even found me another witness, a Washington University professor who had invited me to give speeches.

Having two witnesses made me feel better. But I wanted to make absolutely sure this could end all the trouble. So, on one of my frequent flights from Seattle to San Francisco, I turned on my laptop to search for all occurrences of this document that the lawyers had identified.

I found an email in which I forwarded a presentation by "APCO", a government relations firm doing business with Microsoft. They had presented some materials that I sent to my co-authors writing the paper, "Making It in China." I could not remember this APCO presentation, but the email seemed to indicate that their materials were used. I found and opened the source of this material, and I saw in big letters in front of me: "MICROSOFT CONFIDENTIAL!"

I saw this was why Steve Ballmer was going for the kill. This was the evidence they had! It showed I had taken a Microsoft confidential document, given it to my co-authors to put into another document, and then I released the product to the world, including Google.

Yes the document was given to people IN ADDITION to Google, but it might legally mean that I gave away confidential information to Google.

I could not remember anything about APCO, but the words "MICROSOFT CONFIDENTIAL" required little explanation.

I called my wife from the airport while waiting for my ride.

"It's over," I said.

"How can that be?" She asked.

"I seemed to have taken information from a confidential document and put data from it into another document that I gave to Google."

"What's in the document?"

"Nothing significant. Stories of how other multinational companies succeeded in China. Like how much money HP and IBM made last year."

"So just tell them that this content is public," she suggested

"But I could just tell from the grin on Steve Ballmer's face. He said if the source was Microsoft confidential and Microsoft paid for it, then it cannot be used in public documents. That's also why we failed when we tried to get Microsoft witnesses to admit that VPs could judge what is confidential and what's not. They insisted VPs could not waive the confidentiality label."

"Just relax. There is a way out," she tried to calm me down.

"I don't think so. It's really over. I give up. I will tell Google to blame it on me. They hired the wrong person. They don't need to suffer because of my mistakes."

"Google will defend and support you. Don't give up!"

"I'll come home and search for APCO's document on the Internet. Unless they also released it in the public domain, I'm doomed. I am giving myself 24 hours and then I will need to tell Google to blame it all on me."

At that moment, I felt a chill down my spine. My professional life might be over. My reputation might end. My adventurous job change might turn into doom. My voice was shaking, and I could not control tears coming down my cheeks.

After I went home, I did not find the APCO presentation in the public domain. I called my friend Huang Yong, who had connections with APCO, and asked what he knew. He said to his knowledge Microsoft didn't do much work with APCO, but it was possible that LCA (Legal & Corporate Affairs) did some work. He said he would check for me.

After talking to Huang Yong, I searched my files again. Eventually, I found two copies of the APCO presentation. One was presented by APCO, and the other had a presentation template, "Microsoft Confidential," inserted, probably by a Microsoft person who took it from the APCO presentation to an internal Microsoft presentation.

The APCO presentation was actually NOT Microsoft Confidential! The version I found was embedded in another presentation that was marked Microsoft Confidential, but the APCO presentation was just a business development presentation with general public information. So the "confidential" marking was just a result of a template change. Now, if we could prove Microsoft didn't pay for the data, then we could refute Microsoft's claim.

A few hours later, Huang Yong called back, and said his friend at APCO told him that it was a presentation made to Microsoft as a business development presentation – to try to win business from Microsoft. It was unpaid. He also said the data contained in the presentation were all obtained from public sources.

As I uncovered more and more about the true public nature of this paper, my 24-hour scare was over. But Microsoft didn't know all this, and they would probably start another round of negative rumor spreading to the media...

This was really ugly, uglier than I could have ever imagined.

I needed to call David Drummond, Google's Chief Legal Officer, to find a way to fight back...

## "Gloves Have to Come Off!"

When I reached David Drummond by phone, I quickly spoke my mind, "David, I want to tell you about three things. First, I am willing to do what the company needs me to do. If the company feels it is better to end this case with me leaving, I am willing to do that."

"No, you have done nothing wrong," David responded in a firm voice "We're going to fight this thing."

"OK, then the second thing is: before all this ugliness gets out, can we try to settle?"

"How would we settle?"

"Microsoft should see this is headed for a lose-lose. The newspapers are attacking our integrity, and attacking Microsoft for being outdated."

"Yes, I am afraid it is."

"Why not end this thing and not let both companies suffer in the process?"

"We will try that, both through the outside counsel and I'll call Microsoft myself. But you know, Microsoft is doing this to intimidate other employees, so they are unlikely to settle."

"OK. That brings me to my third point. If it doesn't settle, then we have to fight this hard. Gloves have to come off, if only to make them take notice that we are not just sitting ducks, and if only to give settlement a chance in a few months. And maybe this will help Microsoft see that they have more to lose, and maybe it will cause them to settle."

"Yes, I agree," he said.

A week later, as expected, Microsoft fired the next shot by telling the press that they had new evidence – that I put Microsoft's business secrets into "Making It in China" and gave it to Google.

When a CNET article about industrial espionage became a hit, my heart sank. But only five seconds later, I recovered. I remembered – the gloves had come off, and we were prepared for round three. I called my lawyers, and they released our rebuttal – that the APCO document was all based on public materials, and had been widely published well before I gave a copy to Google. This mud-slinging from Microsoft was all rubbish.

And we prepared more than that – we told the press that the case wasn't about industrial espionage, but about an angry company trying to keep its employees from leaving by scaring them. We gave them an example to show this angry company, an example that would be much bigger than "Making it in China." The story was in the form of a signed declaration by Mark Lucovsky. Mark was a Microsoft fellow and a critical member of the Windows core team. His declaration describes Steve Ballmer's outrageous reaction to his resignation.

In November 2004, Ballmer told Lucovsky, "You can go anywhere, just not to Google."

When Lucovsky refused to comply, Ballmer grabbed a chair and threw it to the other side of the office, breaking glass on a desk top.

Ballmer screamed, "I'm going to fucking bury that guy [Eric Schmidt]! I have done it before, and I will do it again! I'm going to fucking kill Google!"

When this story was released, media attention all turned away from "Making it in China" to Ballmer's anger management.

However, no matter how much harm was done to Microsoft, it would not reduce my suffering and my plight one bit.

Thinking about the upcoming official court hearing, I felt so much pressure that I needed to hear the most supportive voice. I picked up the phone. My wife answered.

After listening to all my worries, she comforted me, "Stop worrying for just a while. Come home for dinner. We will always stand by your side."

That was Sept. 2. After the phone conversation with my wife, I did go home for dinner. Having dinner at home was a luxury during those days, as almost every meal was a working meal with my lawyers.

# Tough Lawyers, Tougher Judge

The Sunday afternoon before the hearing, I met Susan Harriman at the airport and we flew to Seattle together. She and I practiced my answers for the likely questions.

"If their lawyer starts to force me to use adjectives to describe my action, what should I admit?" I asked. "I have not done anything illegal or unethical. But should I concede that some of my actions were unwise, careless, or questionable?"

"You've done nothing wrong," Susan assured me. "Do not start to believe their lies. Admit nothing unless you really felt you did something wrong."

"If I could start over, I would've been more careful and less impatient."

"But have you done anything wrong?"

"No."

"Then keep your head up and tell them that."

I did keep my head up. On Sept 6, right before going to court, I drank two cups of coffee. I wanted to get some exercise but I had forgotten to bring my exercise outfit. So I improvised and jogged around barefooted in my hotel room to boost my energy. When sweat appeared on my forehead, I felt the adrenaline pumping. "This will turn out to be fine," I told myself. "I will win the lawsuit and start a new life!"

The Sept. 6 hearing began with direct examination, in which I was questioned by my own lawyer Brad Keller.

Brad: I know you spend a lot of time on China's education and Chinese students. Is it because of your own background?

Me: Yes, I have a bi-cultural background. I first came to the US at age 11. Since then I've learned a lot from the American way of education and advanced technology of the West. I'd like to bring those back to the East.

Brad: Why do you think your articles are popular in China?

Me: I think the main reason is that they know I'm not doing it for myself or my company, so they trust me. The more they trust me, the more I want to help them.

Brad: Do you have your own educational website?

Me: Yes. I created a website about a year ago. Now it gets at least 20,000 visitors per day. Through the website, I answer about 3,000 students' questions a year.

Brad: Are you writing a book, too?

Me: Yes, it's titled *Be Your Personal Best*. It's meant to inspire young students to develop their best potential.

Brad: Who paid for the expenses of these activities?

Me: I did.

Brad: How did you start writing the article, "Making It in China"?

Me: I started when I felt disappointed at Microsoft's China policy. I wanted to make suggestions. I gathered a lot of case studies as examples. All the materials I used came from the Internet.

After Brad's questions, the Microsoft lawyers started their cross-examination.

Lawyer: Does your article, "Making It in China," contain any business secrets of Microsoft?

Me: Of course not.

Lawyer: Did you use internal information of Microsoft to write this article?

Me: No.

Lawyer: Did you use a business report purchased by Microsoft to write the article?

Me: No.

Lawyer: Where did you get your sources of information then?

Me: Using Google, of course.

The courtroom chuckled. The lawyer started to realize dwelling on the article would get him nowhere. He changed the subject to technology, with an attempt to trick me into admitting I had done search for Microsoft.

Lawyer: Do you know a future product called X Platform?

Me: I wouldn't call X Platform a product. It's still being incubated.

Lawyer: Have you seen X platform being connected with MSN Search to provide search results?

Me: No, X Platform was a partnership with Expedia, who in turn worked with MSN Search.

When that thread got nowhere, the Microsoft lawyer tried to establish that I had spent too much time on Google between May and July.

Most of his questions were of the type, "Did you have this meeting?" And most of my answers were simply to admit them. He meant to accuse me of mixing Microsoft and Google business during the final month before going on my sabbatical. But he couldn't prove anything, other than that I had spent a lot of time thinking about Google in the last two months of my employment.

He then asked me about my sabbatical. I answered his questions, and I noticed that they realized their earlier accusation about my violating a company policy was fabricated, so he did not go into depth here, fearing I would point out their fabrication.

Towards the end, he switched gears. He asked, "Is it true that you have already received a large bonus check from Google?"

"Yes," I said.

"Is it true that you would be paid even if you could not work at Google?"

"Yes."

"Is it true that your stock options would vest even if you could not work at Google?"

"Yes."

"Can you still help the Chinese students through your website during this year?"

"Yes."

"Can you still make an impact this year, and then join Google next year, after your non-compete expires?"

"No, because the press has already printed misleading and false statements about me, which has already hurt my reputation. If this character attack continues, I will not be able to do what Google hired me to do, or help the Chinese students in a year."

The judge listened attentively. It was hard to read his face, but he seemed to be a fair man. He took notes during the section about my sabbatical. At the end he had just one question to test me.

"Dr. Lee, when you went on your sabbatical, your manager asked if you planned to come back."

"Yes"

"What did you say to him?"

"I said I planned to come back. I had to say I was coming back – it was at his staff meeting, and there were a dozen people around. I couldn't possibly say I might not come back due to a Google offer, which I had not received and for all I knew, might not get."

"Was your answer truthful?"

"My plan was to wait for the Google offer, and then consider it, and then decide. I would come back in either case – either to stay or to work with him on a transition."

"So do you consider that truthful?"

This was an impossible question. A "yes" would indicate that I felt it was OK to hold back information, while telling the whole truth was what I had been sworn in to do in court. A "no" would mark me as a dishonest person. This judge was tougher than Microsoft's lawyers.

"It was the truth, but not the whole truth," I calmly replied.

The judge seemed surprised. He nodded and said there were no further questions.

That evening, my sister read the newspaper. She called to tell me how impressed she was with my answer.

The hearing was over. We didn't know the outcome. But we felt good about what we had done. The lawyers told me they had never seen anyone with the kind of composure that I had exhibited. They were confident about the outcome.

At dinner, everyone was relaxed. After weeks of sleepless mad work, this legal team for the first time sat down and talked to each other about something other than Microsoft, Google, and Kai-Fu Lee.

Susan Harriman started a toast and told a joke, "After the hearing today, I ran into the Seattle mayor, who would like to give Kai-Fu the key to the city. He had three reasons. First, to thank Kai-Fu for bringing so much work to the lawyers and journalists in Seattle.
Second, it would be good to keep Kai-Fu from moving to California – we need a smart guy like him. And third, Kai-Fu is influential to talented technologists. With him leaving Seattle, all of Seattle's talented technologists may follow him!"

We all laughed. Then Several lead lawyers thanked their teams.

Finally, I stood up to give my thanks.

"Three months ago, had someone told me I would start the worst three months of my life, I would not have taken that person seriously," I said with a smile. "But had someone told me that in these three months, my only source of fun and joy would be from a group of lawyers, that would've been even more incredulous."

"But that's what happened," I continued, "Because working with you isn't just working with great individuals, but with great individuals who have a heart, who care about me, and who work well as a team."

"And I hope you realize that your hard work hasn't just been about letting Kai-Fu Lee go to work at Google. It's also about the protection of my privacy, the fight for my reputation, the happiness of my family, Google's future business, and most importantly – the 60,000 Microsoft employees whose future freedom depends on your work, and the many more Washington workers whose future freedom rests on the precedence of this case," I concluded with another toast. "To your work. To freedom!"

# The Ruling

Before the September 13 ruling, I enjoyed a rare weekend break with my family in Silicon Valley. We went to the mall. We watched movies. We ate at several great restaurants. We drove to see our old house in Saratoga. We got the kids' favorite – Coldstone ice cream.

Cynthia kept telling me how happy she was to see me home these days. She said, "When you were not home at night, we were all scared!"

I felt guilty about not being there for my wife and daughters while they were adjusting to a new environment. But my wife always said it was OK. It wasn't until much later that she told me she often did not fall asleep on those nights I was away. She was acting strong in front of the kids but weeping all night behind them. She suffered far more than I knew at the time. But she hid it really well at the time. She only talked about uplifting subjects, such as what we would bring to China if I won the lawsuit.

We looked at some furniture that we could ship to China, and told the salesman that on Sept. 13 we would tell them whether or not to proceed with the purchase.

We tried to be optimistic but not over-confident.

Before I flew to Seattle on Sept. 12, my little girl Cynthia gave me a big hug.

"*Bonne chance!*" said Cynthia, who was taking French. *Bonne chance* means "good luck and good courage."

Chimed in her big sister Jennifer, who also gave me a hug.

My wife walked me outside the house to wait for the limousine.

"If this ordeal brings nothing good to me, it has taught me one thing," I tenderly gazed upon her. "That you are the strongest woman I know. Thank you for standing by my side."

She simply smiled and said, "I know right from wrong."

"If we lose, I will retire to spend more time with you."

"You won't lose," she said with utmost confidence. "You will win back your reputation. You will do great things for the world!"

I hoped she was right, and I knew I had done nothing wrong, but I still worried at the same time. On Sept. 13, 2005 at 9 a.m., I nervously walked into court, accompanied by two of my lawyers, John Keker and Ragesh Tangri. I sat down with my heart in my throat.

The air was still, almost suffocating. Time seemed to slow down. A second, two seconds...It seemed to be taking forever for the judge to show up.

He finally appeared. We all stood up when he came up, and sat down when he took his position. I held my breath, getting myself ready for the announcement that would determine my future.

When the ruling was announced, I almost couldn't believe my ears---the Washington State Court approved all of Google's requests about my employment!

Based on Google's requests and the ruling, I could immediately start working for Google but would not be involved with search or language-related projects. I could go ahead to establish Google China and recruit talents there. I could also take charge of Google's branch location, public relations and government relations in China.

The ruling specified that I could provide Google technology and business strategies as long as I would not use Microsoft's confidential information.

The judge said my job duties for Microsoft Research China prior to August 2000 should not be considered in this case because the non-compete agreement I had signed for that position had expired in 2000.

As for the non-compete agreement I signed after August 2000, the court would decide on whether it should be upheld in another hearing in January.

"We won the first battle!" exclaimed John. He excitedly gave me a hug.

Ragesh also shook my hand to congratulate me.

Reporters soon gathered around me with microphones pointed to me, asking for my comments on the ruling.

"I'm happy I can finally go to work in China," I said with simplicity and sincerity. "I'm grateful to my lawyers. It's because of their tireless effort that I get to realize my dream in China. I'm ready to start work at Google China!"

As soon as I finished my statement, I overheard a Microsoft legal vice president telling reporters, "Google just hired someone who can't work on search! Google is hiring the most expensive human resources manager in history!"

I was afraid his words would mislead the media and start rumors again, so I distracted the reporters around him by raising my arms and making two Vs with my fingers. All those reporters immediately turned to me with their cameras focused on me. Later, the most eye-catching part of almost every news story on the ruling turned out to be my Vs in the attached photo. Given most editors' common practice to choose titles that would match photos, the titles of those stories were favorable, too. A picture was indeed worth a thousand words in this situation.

After the ruling, I couldn't wait to fly home. My daughters ran to the door to welcome me back. Cynthia jumped up onto me. I raised her up and held her in my arms.

"Daddy, you said I could get a dog if we go to China," she reminded me in her sweetest voice. "Does that mean I can get a dog soon?"

"Sure!" I kissed her on the cheek. "When we go to Beijing, you can get as many dogs as you want!"

That night I turned on the computer and saw all the Internet reports on the outcome of the lawsuit, my regained reputation, and the plans of Google China. I also received a Chinese poem from someone I had hired at Microsoft Research Asia celebrating the occasion.

The poem was acrostic with my name embedded in it, and could be roughly translated to:

*With all the dust settled and the air cleared,*
You will move forward, traveling light and reaching high
*May you create a bright future*
As you return to China for a new venture!

Smiling at the poem, I finally felt relieved. I picked up the phone, called the furniture store I had visited with my family, and told the sales person, " I'd like to place an order."

## Reconciling with History

When I landed in Beijing on Sept. 17, 2005, the ground looked wet, but the sky was clear. The weather seemed to reflect my heart's relief from the worst storm of my life as well as my sunny outlook on the prospect of my new career.

Google held a press conference in Beijing to announce the establishment of Google China. I thought the media would ask a lot of questions about the lawsuit. But they all focused on recruitment, company location and product strategies. They didn't even mention a word about the lawsuit!

Then I realized the lawsuit was already over in their eyes, though in fact it was not.

In November 2005, I flew back to Seattle for one more deposition. I squeezed time from my busy work schedule to prepare for the lawsuit. My former student Alan Guo also did a lot of research for me. He looked through legal precedents and found that the non-compete agreement couldn't sustain without the company's offer of additional compensation to the employee. I was excited about his discovery and looked forward to using it as favorable evidence. But unexpectedly, the Microsoft lawyer only asked me two technical questions in the deposition.

After the deposition, my lawyer Ragesh Tangri came up to me and said, "Kai-Fu, we settled. Microsoft withdrew their case. The lawsuit is over."

"It's over?" I was stunned. "All of a sudden it's over?"

"Yes," he confirmed. "They quit because they knew they were not going to win."

"If so, why did they ask me those technical questions?" I still found it hard to believe.

"That was just to clarify what you did at Microsoft and put it in the settlement," he answered and handed me a piece of paper. "Look, this is the settlement."

I browsed through it and found a timetable of when I could start doing what type of work for Google.

"Such favorable conditions!" I expressed my surprise. "How did you get these?"

"Actually, everyone was tired," Ragesh explained. "They dragged it on for two more months only to save face. But they knew the next hearing would be next month and they had no chance to win, so they finally agreed to settle."

"That's wonderful!" I smiled. "But if I told everyone about this agreement, wouldn't that mean they lost?"

"No, you can't," said Ragesh. "Look at the last clause."

I took a look. The last clause mandated both sides to keep the agreement confidential.

"If we don't publicize this," I hesitated. "Will people think we paid them off for violating the non-compete agreement, or that I'm still restricted by the non-compete agreement?"

"No, Kai-Fu," Ragesh assured me. "When the financial statements of the next quarter come out, the media will see we didn't pay them off, and people need only look at Google China's progress to see your work areas have been expanded. Everybody will know then."

"OK," I shrugged. "The media don't seem to care about this case anymore anyway."

It ended just like that. The so-called "lawsuit of the century" that once caused worldwide media sensation and cost tens of millions of dollars quietly became history.

Today I wonder how many of the general public still remember the lawsuit. People tend to have short memories of past headline stories. As they watch breaking news every day, they need to empty some storage space in their brains, like deleting obsolete computer files, in order to take new information. The lawsuit therefore may have been forgotten, or soon it will be.

To me, however, the lawsuit has become a lifetime experience I can never forget.

For quite a long period of time, the lawsuit was like a deep wound in me that no one could touch. The wound would bleed every time I recalled how Microsoft had once damaged my reputation, how some of my protégés had participated in the company's plot against me despite all that I had done for them, or how some friends had distanced me...

But at the same time I would think of how other friends showed their unfaltering trust in me, how Xu Xiaoping agreed to be a character witness for someone he had only met once, how Mary Hoisington was willing to come out of her retirement to testify for me, how my friend Huang Yong became my source of ideas and encouragement, and how my protégé Alan Guo spent countless hours collecting favorable evidence for me...

I would also remind myself how my wife always gave me strong support without letting me see any of her fears or tears, and how my daughters cheered me up while I was the most down...

I still had a lot after losing a lot, and the lawsuit taught me to cherish what I had.

My wound gradually healed. I began to smile more and more. One day in my first year with Google China, one of the employees said to me, "You seem fearless. You can tell jokes to cheer us up even when we face the biggest challenge." Then another added, "Everyone says this is a characteristic of your unique leadership."

As I contemplated their remarks, I realized the fearlessness they were talking about was something I had gained from the lawsuit. After being almost beaten down, I had bounced back, becoming more resilient than ever.

Now looking back, I no longer feel resentful toward Steve Ballmer for his taking me to court. I've decided to hold no grudge against Microsoft. Instead, I'll keep fond memories of those working nights with tireless team members at Microsoft Research Asia, splendid summers in Seattle, pleasant conversations at the Microsoft annual picnics, and every inspiring minute I discussed technologies or strategies with Bill Gates.

I remember a passage from a book that essentially says, "Forgiving is a kind of salvation. It saves everyone from the shadow of a past disaster. This is the only way to reconcile with history."

I've reconciled with my lawsuit history.

*My choice of smiling at lessons learned*
*from the so-called "lawsuit of the century"*

## CHAPTER 11

# Taking Google to China

When I first drove into Google's headquarters on Sept. 14, 2005, I felt as if entering an amusement park. There was a gigantic dinosaur skeleton in the central courtyard. As soon as I stepped into one of the four office buildings, I saw a giant model of space shuttle, and an antique telephone booth in a corner. It was hard to believe I had once worked in this complex for SGI.

The four buildings of the Google headquarters were initially built for SGI during the company's heyday. Then-SGI-CEO Ed MacCracken hired the most famous architects to design the most impressive office buildings of Silicon Valley. The construction cost twice as much as the average price per square foot. The ceilings of the buildings are also much higher than normal. The highest one is more than 30 feet from the ground.

The shiny glass walls of the four buildings were dramatic, as suitable for the Hollywood-oriented SGI as for the Never-never-land of Google. But in the interior, Google made drastic changes after purchasing the complex for $319 million.

In the Google headquarters, there are not only offices but also video game machines, ping-pong tables, a gym, a shower room, a laundry room and many massage rooms with licensed therapists in them. There are even two roof-top swimming pools.

There is a story behind the swimming pools. One day, some engineers of AdSense, a Google team working on the technology of displaying ads, said Google had everything but a swimming pool. Co-founder Sergey Brin responded, "OK, if you can complete your project on schedule, I'll build a swimming pool for you."

The engineers did submit their project on time. They went to Sergey's office to ask for the swimming pool. Sergey said OK. Then the next morning the engineers saw a plastic kiddie pool in front of their office. Sergey and the other co-founder Larry Page appeared in their swimming trunks and said, "Here's your swimming pool!"

Of course that was just a joke. The two founders later had real swimming pools built. The two roof-top pools are small due to space limitations, but they have strong currents that can keep you swimming in-place as if running or walking on a treadmill.

Google provides everything everyone may possibly need all day so no one has to step out of the office for anything. It's the company's way of making employees naturally spend more time at work.

The company has broken the there-is-no-free-lunch convention for the same reason. Giving employees free meals on site eliminates their need of going out for lunch, and the saved time will turn into higher productivity.

To keep employees on site during their lunch break, Google meals are not only free but also fabulous. The chefs all came from top restaurants. The first executive chef at Google, Charles Ayers, was especially a legend.

Ayers came on board early as Google's 56th employee. He created a 135-member chef team at Google, using all organic ingredients to cook a variety of meals, including international styles such as Chinese, Italian, Japanese, Korean, Mexican, and Thai cuisines, plus vegetarian and Kosher dishes. To make sure the team cater to every Googler's taste buds, Ayers often conducted surveys.

His effort paid off. Charlie's Place, the first cafeteria at Google, became more famous than most restaurants in Silicon Valley. Googlers were proud to invite their family members and friends there. Every non-Googler who had a chance to eat there would keep raving about it afterwards.

When I joined Google, Ayers had resigned with all the money he had made from Google's stock to open his own restaurant. But the largest cafeteria at Google was still called Charlie's Place, where abalone, lobsters, Alaskan king crabs, fresh fruit juices and homemade ice cream were served.

I saw two other cafeterias. One provided Asian food including Chinese dim sum and Japanese sushi. The other was a special salad bar with all kinds of dressings freshly made from raspberries, peaches, mangoes and coconuts. There was also a larger variety of juices than normally seen at restaurants. I had some wheat grass juice there. Although its taste didn't really appeal to me, I appreciated its high amount of nutrients.

Later I learned that Google opened more cafeterias. Today there are more than 10 of them. It's Google's policy to have food within 100 steps from everyone's desk. Besides the cafeterias, Google also has many "mini-kitchens", which are really snack stands. Employees can fetch snacks from those mini-kitchens any time, just as they can go to a Tech Stop at Google to pick up all kinds of electronic supplies and gadgets.

Google is doing everything to pamper employees, to make them feel even more comfortable than at home, and the company allows employees to bring pets from home. I saw numerous little dogs in the office buildings. To my surprise, there was also a huge animal. I thought it was a horse when I first spotted it from a distance. Then someone told me it was vice president Urs Holzle's leonberger.

"It's Google's first dog!" said the Googler who was talking to me.

What impressed me more than the casual environment was Google's egalitarian culture. I saw everyone's office in the same size and everyone being treated as equals instead of supervisors and subordinates.

My former student Alan Guo felt the same way when he was a product manager at Google. One day he went into a conference room for a meeting, but he didn't see the people he was supposed to meet. Instead CEO Eric Schmidt and the two Google founders walked in.

"Uh-Oh!" Alan thought. "I must have entered the wrong room!"

However, none of the three leaders asked him why he was there. They just started talking about Google's European business strategies. They made Alan feel he had the right to be there to listen if he chose to.

Google is a company that allows employees to make mistakes, as it sees mistakes as learning experiences. Vice president Sheryl Sandlberg said that she had once made a serious mistake costing Google millions of dollars. But Larry Page didn't blame her at all. Instead he said, "If we didn't make any mistakes, it would mean that we hadn't taken enough chances!"

I was more than happy to be part of this almost-too-good-to-be-true company. Google held a welcome party for me. It was a big party with hundreds of people, and many of them were my previous colleagues at SGI or Microsoft!

I opened my speech with a typical American joke, "Today is the happiest day of my life, because it's the first day I no longer have to work with lawyers!"

As people laughed, I continued, "Actually I appreciate my lawyers very much. It was their tireless effort that finally brought me to work here. What really makes me happy today is I'm going to work with you guys, the brightest engineers in the world! I can't tell you how excited I am about it!"

At that moment I recalled the fireworks I had seen on the fourth of July, right before leaving Microsoft. I found Google as breathtaking as those fireworks. It made me feel 10 years younger.

During those few days I worked in the headquarters, some Googlers I didn't know would say to others, "I had a Kai-Fu Lee sighting today!" When I overhead such a description, I smiled, feeling as if being a celebrity.

## Tricky Exam Questions

One of my days at the headquarters, a gray-haired man in a three-piece suit walked into my office. He said, "Hello, Kai-Fu! I'm your new roommate Vinton Cerf."

Vinton Cerf! The father of the Internet! My goodness! It seemed to me that Google had recruited every brilliant computer engineer I knew by name!

*Chatting with Vinton Cerf (right) at Google headquarters in 2005*

While marveling at Google's extremely proactive recruitment, I was aware that it was only the cream of the crop Google was seeking. Google once had an exam published in a few major magazines in 2004, intended for people to test their own "Google aptitude."

The exam only consisted of 21 questions, but most of the test takers would spend hours on it, still with some of the questions unanswered.

The exam contained real math questions. One of them read, "Paint a 20-surface polyhedron with three colors and make each surface single-hued. How many combinations will there be?"

The exam also had unconventional questions, such as "What's the most beautiful math formula in the world?"

In addition, there were questions asking test takers to display their creativity. One question just asked the test taker to fill in the blank with something impressive. Another requested the test taker to redesign the visual presentation of the exam.

Google recruiters told me that the company had received answers to the exam questions from tens of thousands of people. Some of them were mathematicians and professors with exceptionally high IQs. Others were puzzle-solving experts. These people were not interested in working for Google. They just wanted to test their abilities."

In addition to the exam, Google also posted {first 10-digit prime found in the consecutive digits of e}.com on advertising boards near Stanford University and MIT. Those who solved the question would reach a website where there were more computing questions. Eventually, those who answered all the questions right would be taken to a Google recruiting site, where they could apply for a job.

I knew I was supposed to bring this recruitment style to China.

In order to instill a true Google culture in Google China, I needed to recruit Chinese engineers from the headquarters. I was afraid their supervisors might be unhappy about them being taken away. But all the engineers said, "Don't worry. This is for the good of the entire company. Plus at Google, it is always the engineer's choice! Our bosses will all support!"

Since I had bad memories of departments fighting against one another for talents and resources at my previous workplaces, I greatly appreciated Google's different culture.

Google let me bring two product managers to China. One of them, Alan Guo, had been my student at Microsoft Research China and had earned an MBA from Stanford University. Alan is a quick thinker. He frequently comes up with solutions while others haven't figured out where the problem is.

The other product manager, Ke Yu, is smart but shy. He followed his parents to Brazil as a child and went to high school there. Then he went to college at UT Austin and graduate school at Stanford. When people learn his background, they often say, "Wow! You were born in China and then grew up in Brazil and America. You must know how to play ping pong, soccer and basketball!" To such comments, he always responds, "Well, unfortunately, I play soccer like a Chinese, basketball like a Brazilian, and ping pong like an American!"

Joining the two product managers were three Google engineers, Ben Luk, Hui-Can Zhu, and Hong Zhou.

Ben was born in Shanghai and grew up in Hong Kong. He went to college at Cornell University and graduate school at Stanford. Hui-Can, a Cal Ph. D., had the highest seniority among ethnically Chinese engineers at Google. Hong joined Google right after finishing her Ph. D. at Stanford. She went to college at the age of 15, and was a "poster child" in Google's recruiting announcements.

I also recruited a marketing talent named Ning Tao from IBM. Ning had been a product manager at Microsoft and had masterminded the design of a well-received Chinese launch of Windows 98. Ning was going to take charge of operations at Google China.

The seven of us formed the first team of Google China. Our first task was recruiting. I flew to a different city in China every day or every other day. Alan and Ke would go to every city ahead of me to lay ground work for me.

At our first stop, Xian Electronics Technology University, the students' enthusiasm far exceeded our expectations. The auditorium with a capacity of 3,000 almost exploded, and there were about 3,000 more outside the doors, as the university broadcast my speech to those outdoors. I heard many of them had come from other cities.

For my later speeches, some universities distributed tickets to students on a first-come, first-serve basis and only allowed those with tickets to get in. But some students produced their pirate versions that looked just like real tickets, with which they took the real ticket holders' seats. Then the real ticket holders arrived. In such situations, verbal fights inevitably broke out.

Fortunately, all the arguing students would stop when they saw me appear on stage.

Only once in Shanghai, when I walked onto the stage, they didn't immediately quiet down. Instead I heard many Ahs from the audience. "What's going on?" I silently asked myself, feeling puzzled.

At this moment, a student in the audience loudly answered my unspoken question. He said, "Dr. Lee, you look a lot skinnier!"

"Oh! I see," I said to myself. These were students I had met before. But they hadn't seen me for a long time.

I smiled and said, "So, someone just said I got skinnier. Let me ask everyone, who would like to know my secret to losing weight? Please raise your hand!"

Many students raised their hands.

"OK," I nodded. "You all want to know, so here's my secret: If you want to lose weight quickly, just let Microsoft sue you!"

Everyone burst laughing. I laughed, too. But at the same time tears almost rushed to my eyes. I thought of how dedicated I had been to advising Chinese college students since 2000, and how much I had been once afraid of losing the students' trust for the Microsoft lawsuit. What happened in front of my eyes told me the lawsuit was really over. I had really gained my reputation back.

After almost losing the mentor's role, I cherished it more than ever and vowed to be even more devoted to it.

From mid-September to mid-October in 2005, my team and I flew to numerous Chinese cities. Almost every day of that month I got up at 5 a.m., left the hotel at 6 a.m., boarded a flight at 8 a.m., entered a university campus at 11 a.m., gave speeches at various universities until late at night, and checked into my hotel at 11 p.m.

My team members saw my tired looks when they picked me up at the airport. Then they would be amazed to see how energetic I appeared in front of the college students. They said there was magic in me. Later they followed a journalist's description to call me a "steel man" for my invincibility to the exhausting schedule.

My speeches always preceded written exams we gave to interested college students. My team would look through the exams right after collecting them. There were so many exams to go over that the team members often didn't have time to go out for dinner. Then I would buy takeout food for them.

Once I bought them 50 boxes of various snacks, and they ate them up after working all night!

When the recruitment began to show exciting results, I realized too frequent flying had left me a health problem. The right side of my back was as hard as a rock! Even massage therapists couldn't do anything about it. I jokingly said I had truly become a "steel man!"

More than two years later, I finally found a massage therapist who was able to move muscles on my back. I wondered why he had stronger hands than all the others until he told me that he had been a national champion of track and field athletics.

# The Establishment of Google China

In September 2005, Google China's co-president Johnny Chou arrived. Before he started his job, we had a talk at a coffee shop. Johnny directly said, "Kai-Fu, I believe neither of us joined Google China for money, so the most important thing is for the company to succeed. No matter how we are going to get along, we have to stand by each other."

I absolutely agreed. That day we promised each other to discuss our different opinions in private in case of any disagreement. This way Google China would have only one voice in public.

We decided to share an office in order to make it more convenient for us to discuss anything any time. This was nothing unusual at Google, but many Chinese reporters were surprised to see us sitting in the same office.

Johnny and I cooperated harmoniously. He took charge of Google China's sales and partnership while I was responsible for government relations, public relations, marketing, and, of course, products and technology.

*Sharing an office with Johnny Chou in 2005*

Google China planned to hire 50 engineers for the first year. I told the media that I was going to personally mentor these 50 engineers, and it would be the last year when the company would be small enough for me to do that. The news soon spread out and attracted thousands of applicants.

Every time we gave a written exam, there were more than 1,000 test takers. But only 50 to 100 of them would make it to a phone interview. Then even fewer would go to an on-site interview. Eventually only one out of every thousand applicants would be hired.

Those with years of work experience didn't have to take a written exam. But they had to go through phone interviews and on-site interviews. Sometimes an applicant might be on the border line. In this case we would ask the headquarters to conduct more phone interviews before making our decision. There was an applicant who had to take 12 phone interviews for this reason.

Unfortunately, the applicant was rejected in the end. His wife wrote me a seven-page long email, in which she said, "My husband wants to join Google very much. Every time he received a call from the Google headquarters, he got all nervous. It was hard for him to find a quiet place to talk because we have children and live with his parents. Sometimes he had to do the phone interviews in the bathroom to make sure there was no noise. Sometimes the headquarters called in the middle of the night. He would jump out of the bed right away to answer. But after going through so much, he received a no from you. The rejection letter shattered his dreams…"

Her email made me feel terrible. However, I was unable to do anything to help. The company considered rejections final. All I could do was to improve interview methods for future applicants. We decided to do no more than eight interviews per applicant and never to call an applicant after 10 p.m.

Some applicants with outstanding performance in technology failed our interviews because of their unpleasant personalities. One of them was a famous professor teaching in Asia. It took me quite a bit of effort to talk him into an interview with us. But Google China's other interviewers told me he was totally arrogant. They said he wouldn't mix well with our egalitarian company culture, so we didn't hire him.

Another such case was a student who received nearly perfect scores on all the sections of our written exam. But when he saw one of our interviewers, Niniane Wang, he inappropriately commented, "I can't believe you are so young! You look only about 18, and you are the interviewer?" Niniane was a little annoyed but let it slide. What really upset her was the applicant's reaction when she told him he had answered a question wrong. He protested, "Do you think you are better than me? Why don't you let me test you with a question?"

The other interviewers all agreed with Niniane that we couldn't hire someone with that kind of attitude.

Every single one of our interviewers had the right to veto the panel's decision. One panelist, a senior engineer from the headquarters almost used his veto right when reading an applicant's college transcripts. The applicant did extremely well on our exam and at our interviews. But he only received barely passing scores on some computer science subjects in college. The panelist said, "Grades reflect a sense of responsibility. What if he is irresponsible and brings the same habit here?"

All the other panelists saw great potential in the applicant, so they looked at me as if asking me to override his veto. But I respected his veto right, so I said, "Why don't you give the student a call? See what he says about his grades. If he can't convince you, we'll just turn him down."

The senior engineer agreed. A few days later, he sent me an email, in which he said, "Let's hire him! The student said his school had a very bad computer science program. The professors didn't know much and the exams didn't have much to do with programming. I never realized some Chinese schools are so poor. And what a smart lad he is to have learned so much on his own, and to have studied just enough to slide by!"

Some Chinese students displayed amazing talent during their interviews. One of them especially impressed us after answering five questions quickly and perfectly. When facing the sixth question, he said, "I'm not completely sure about the answer, but I have three approaches that may help me find the answer." Then he expressed his three approaches clearly. We knew immediately he was exactly what we were looking for, a creative and practical engineer!

But when we gave him an offer, he hesitated. He said his mother had never heard of Google and wanted him to work for a more established international company in China like Microsoft or Intel. In the meantime his girlfriend encouraged him to do graduate studies abroad.

To recruit this rare talent, I wrote a long letter to his mother. I also took him, his mother and his girlfriend out for dinner and explained how many learning opportunities Google would offer him. Finally, I convinced his loved ones to let him join Google China.

Besides interviewing, we also took recommendation letters into serious consideration. One recommendation letter came from one of my former colleagues at SGI. The letter unconventionally said, "This applicant is as good as 10 engineers. Whoever doesn't hire him is a fool!"

We soon found out during an interview this applicant was indeed a genius. Of course we didn't let ourselves become fools. We hired him right away.

By January 2006, we had recruited more than 50 engineers.

Our next step was to hire an executive chef with working experience at a five-star hotel, following the headquarters' tradition. We were looking for a Chinese version of Charles Ayers.

To find an executive chef who could cater to everyone's different taste buds, we had every department of Google China send a representative to form a committee in charge of interviewing candidates. Two finalists would each cook a meal for the committee, and then the committee would vote.

We happened to have a pajama party at Google China when the two finalists came to cook. The food committee members entered the dining hall in their pajamas and slippers as if having a Halloween party.

The finalist cooking lunch was a dark-skinned, middle-aged rotund man from Beijing. He had worked as a chef on many luxurious cruises, so he knew how to cook Western food exceptionally well.

The one cooking dinner was a small-framed young man coming from a five-star hotel in Qingdao, a city in eastern China. When I asked him what dishes were his specialties during an interview, his answer quite impressed me.

He said, "I wouldn't call any of my dishes my best because I consider all of them my specialties. All my dishes come from my improvement on the recipes I've collected. I believe a good chef has to know how to reinvent recipes and create new dishes in order to cater to thousands of different customers."

"That sounds great!" I gave him some encouragement before asking the next question. "Do you know how to make Western food? Are you communicative in English?"

"I have more than 1,000 Western recipes in my notebook computer," he said. "I've cooked a lot of Western food, and I've taken cooking classes in Paris. So, I can read not only English recipes but also French recipes."

I nodded, and asked one more question, "Why do you think you are the right person for this job?"

He replied, "I've read a lot about Google. I found myself a lot like Charles Ayers. I enjoy making people happy through my food, so I always communicate with customers, to understand their needs and preferences. For example, in order to make Muslim food right, I've talked to many Muslims. If Google has Muslim employees, I guarantee that I can offer them the most authentic Muslim food!"

My sixth sense told me that this man named Rohnsin Xue was the perfect candidate for our executive chef's position. As soon as he became one of the two finalists, I began looking forward to tasting his food.

Rohnsin cooked abalone, lobster, crab legs and lots of other delicacies for the contest. All the food committee members couldn't stop eating until feeling more than stuffed. He won an overwhelming majority of votes.

From then on, Rohnsin walked around the dining hall in his chef uniform every day to greet us, asking what we felt like eating. He also conducted surveys on line and changed his menus accordingly. When we particularly loved a dish, we would have as much of it as we could right then because we knew it wouldn't reappear for quite some time. Rohnsin would never repeat a dish within the same week.

In the meantime, Rohnsin never stopped collecting recipes. When he heard me rave about my mother's beef noodle soup, he immediately decided to introduce it to Google China's cafeteria. In the beginning he didn't get it exactly right because he didn't know the importance of adding *suan-cai* (Chinese sour kraut) or the difference between the shank from a cow's front legs and rear legs. But after continuous experimenting, he finally grasped the best of my mother's secret recipe and incorporated his own creation into it. He increased the proportion of peppercorn and added other spices. In the end his beef noodle soup was even better than my mother's!

Rohnsin also improvised frequently. During the Olympic games in summer 2008, he made a dessert in the shape of a bird's nest in front of international guests on the spot and blew everyone away. He also impressed the entire staff of the headquarters with his ice sculptures when he attended training courses there.

*Google China's cafeteria with gourmet food and a creative executive chef, Rohsin Xue*

Everyone said Google China successfully transplanted the headquarters' food culture while doing a thorough job in localization. We also had the same kinds of facilities available as in the headquarters. All the equipment in our gym came from the headquarters. We offered professional back massage and foot massage like the headquarters as well.

Most importantly, we duplicated the headquarters' egalitarian culture at Google China. Like the headquarters, we held a TGIF (Thank God It's Friday) meeting every week and encouraged all the staff members to express their opinions.

Since our Chinese engineers were raised in the hierarchal Confucian culture, they were cautious around the boss and didn't have the habit of speaking up at meetings. To loosen them up, I played a "Dance Revolution" machine to perform a little dancing in front of them at a

TGIF meeting. They couldn't help but admire my high score yet laugh at my clumsy moves. Then they felt more and more at ease about expressing themselves to me.

However, some employees became too relaxed and sometimes arrived at our management retreat late. I wanted to send a clear message them, but in a humorous way, so I said, "Next time whoever comes late will do belly dancing for us. Does everyone agree?"

"Yes!" "Oh yes!" Everyone seemed excited.

I wasn't joking. From then on, those arriving late did have to perform belly dancing. One time Alan Guo was late, and he had to do it. He was a good sport, and put up quite a show. All the employees laughed non-stop when they saw him trying very hard to come up with some moves.

After that, no one was ever late again.

## Weathering Storms

On Jan. 25, 2006, Google launched google.cn, a website specifically designed for China's Google users. Google doesn't necessarily establish a local website along with its branch in every country, because that means the company has to move servers to the country and abide by local laws, including search result censorship. But when CEO Eric Schmidt asked for my opinion, I said, "It's necessary to set up a website for Google China. You can ask all the Chinese engineers at Google to see if they think the same way." Then Eric told me he had asked all of them and indeed they all agreed with me.

The headquarters accepted our suggestion. However, as soon as google.cn was announced, Western media began to criticize Google for going along with China's censorship to display filtered search results, which contradicted Google's principle in spearheading the freedom of speech.

Many engineers at the headquarters questioned the decision at the same time. They thought google.cn would hurt Google's image as the world's number one brand in search. I felt an urgent need to communicate with them, so I put aside a lot of work and made a special trip to the headquarters, where I held a meeting that welcomed everyone interested in the issue to attend. More than 300 headquarters employees showed up.

I spent a lot of time and effort to help them understand why it was necessary to abide by Chinese laws in China. Finally, I saw most of them nodding.

But that was not the end of it. Pressured by media, the United States Congress decided to hold a hearing for controversial high tech companies such as Microsoft, Cisco, Yahoo! and Google to express themselves on sensitive topics. That meant Google would have to explain its China policy to all Americans.

I happened to be at the headquarters when Google was asked to attend the hearing. Sergey Brin said to me, "Don't worry! Even if we decide to withdraw google.cn, we won't let go any of the talents you've recruited. And even if we do eventually back out, we will keep an R & D center in China!"

I thanked him for this promise, but at the same time I was surprised that he was actually considering the possibility of withdrawing google.cn!

On Feb. 14, 2006, Google's top management held a meeting to discuss google.cn. I was already back in China then. But I couldn't stop thinking about what decision would come out of their meeting.

The meeting went on until after 11 p.m. in California time. Then Google's representatives took a red-eye flight to Washington DC for the hearing. Sergey made an international phone call from there to my Beijing office. He said, "Kai-Fu, we talked until very late last night.

It was the first time Eric, Larry and I couldn't reach an agreement. But the hearing is today, so we have to tell everyone our final decision. We are still going to China!"

I felt relieved! The next day I told all the employees of Google China about Sergey's phone call, and I assured them, "No matter what, we won't lay anyone off, because Google sees talents as the company's most precious asset!"

Just when all the stir about the Congress hearing gradually quieted down, another media storm broke out in China. On Feb. 21, 2006, a headline story titled, "Why Google Climbed over the Wall into China," caused a sensation. The article accused Google China of operating without an Internet license (known as "ICP") and pointed out the license number appearing at google.cn belonged to another website.

Google, a highly esteemed company with "Don't be evil" as its motto, was operating illegally in China? The news shocked everyone, including us working at Google China.

As a matter of fact, all the international companies in China then had come in with borrowed Internet licenses. Although a new Chinese law passed in 2002 mandated international Internet companies to establish joint ventures with local companies, those coming after 2002, such as Yahoo!, eBay, and Amazon, still followed the old routine to just borrow licenses.

When we asked Chinese lawyers about what Google China was supposed to do, all the lawyers suggested borrowing a license because it would be much quicker than applying for a joint venture. That was why we began to operate on a borrowed Internet license.

Given Google China's high profile, however, we realized we couldn't tell the media we were just doing what everyone else did. We started negotiating with the Chinese government to explore the possibilities of setting up a joint venture.

Before the dust settled, rumors were everywhere in the air. I had to reassure all the employees again and again that there wouldn't be a layoff even in the worst case. I also promised to keep them posted on everything I heard from the Chinese government.

The following two months felt longer than two years. Finally, we reached an agreement with the Chinese government. Google China agreed to apply for a joint venture immediately, and before the approval of the application, Google China was allowed to continue its daily operations.

While weathering the two big storms in early 2006, Google China lost a great timing in securing market share. At the time most Google users in China still went for the Chinese language service of google.com, which sometimes had connection problems because of Chinese government blockage. We wanted all the users who typed Chinese on google.com would be immediately directed to our new google.cn, but that couldn't take place until our joint venture was established.

In the meantime, our staff had a little disagreement on the directions of our product development. Were we supposed to focus on the quality of Chinese language search? Or were we going to develop other cool types of products? Debates about those issues consumed our energy.

Some people asked me how I felt about all the adversaries Google China faced. I told them I was optimistic. Even though Google China in its beginning stage didn't run as smoothly as I had imagined, I believed in my excellent team. I was confident that Google China would take off as soon as all the logistic problems were squared away.

In April 2006, in order for Chinese users to feel closer to Google, we phonetically translated the company's name into two Chinese words with a nice meaning, "a song from the valley." This was the first time Google had its name translated into another language.

In the classy lobby of Beijing Hotel, then-CEO Eric Schmidt and I put together pieces of a huge puzzle to display the two characters of Google's new Chinese name. As soon as we finished the job, all the audience members cheerfully applauded as if they had already seen a bright future of Google China.

## Perfecting Google's Chinese Language Search

Google's Chinese language search began in 2000 with a Chinese language service on google.com. A small team of only five engineers at the headquarters worked on it, with all its servers in America. In the meantime, several Chinese search engines emerged in China. Their knowledge of the local culture and laws put them in an advantageous position over Google's Chinese language service.

When Google announced the establishment of Google China, a video parody called "I Know You Don't Know" immediately became widely circulated on the Chinese web. The short film was basically making fun of Google, implying it would be impossible for a foreign company to do well on Chinese language search.

Under such circumstances, I was certain that our first priority was to produce high-quality Chinese language search. We had to put everything else aside to focus on Chinese language search technology.

The director of our product department, Ke Yu, agreed with me. He said, "Search is Google's core technology. If we succeed in search but fail in everything else, we can still be considered successful. But if we do well on video, communities, and maps but give up search, then we will be looked upon as a failure!"

His insightful remarks made those engineers who suggested making other cool products right away change their mind.

Then-CEO Eric Schmidt fully supported this plan. He said, "Search is Google's secret to success. It also lays the foundation for advertising."

*Accompanying Eric Schmidt on his China visit in 2006*

I passed his words to all the engineers at Google China, and I said, "It's not that we are never going to produce the coolest, the most eye-catching products. We will, but that'll be in the future, not right now. For now let's all do our best to concentrate on search!"

In Chinese language search, what we needed to do first was to improve the grouping of Chinese words. As mentioned in Chapter 8, a Chinese word may or may not carry a meaning on its own, and two or more Chinese words are frequently grouped together to serve as one unit of meaning. When someone types a phrase in Google's Chinese search box, it's necessary for our system to dissect the phrase into word groups in order to figure out what the user is seeking.

For instance, if the user typed *"niu rou mian shi pu"* in our Chinese language search box, our system would group *niu, rou,* and *mian* together as the combination of these three words means "beef noodles," and at the same time it would connect *shi* and *pu* for the two words to signify "recipe." Then our search engine would find all kinds of beef noodles recipes for the user. But how would our system know to group the word *mian* with the two words preceding it instead of the two words following it? That came from our hard work.

From late 2006 to mid 2007, we worked tirelessly on Chinese word grouping. We also must give credit to Jun Liu, director of Google's Search Efforts in China, for his brilliant suggestion of applying my statistic model of speech recognition to Chinese word grouping in search. Given Chinese words being all one-syllable, it was quite manageable to treat them like sounds. Using statistics helped us tremendously.

We created four metrics to evaluate our progress: 1) relevance of the search results; 2) the size of Chinese web page indexes; 3) updating speed; 4) recognition of spam websites.

We checked the four metrics every day and compared them with what our competitors did.

To increase our understanding of users, we hired several psychologists to conduct surveys and experiments. We also invited some users to browse the web in a room of our office building and had our computers record their behavior. Then we saw some differences between Chinese and American Internet users. While Americans tend to leave the search result page right after finding what they need, Chinese users generally spend more time on the search result page to explore all kinds of information.

We also found Chinese users to generally type fewer key words than their American counterparts. We believed it was partially because it would take a longer time to type Chinese. For the user's convenience, we began to offer relevant suggestions for them to choose after they started typing. This feature, called Google Suggest, was first launched in China and then adopted worldwide.

We tried to please our users in every way. We even adjusted the layout of our web pages, trying every font size and spacing gap, to what they liked the most.

I thought about how to do better on Google's Chinese language search every minute of my waking hours, and I dreamed about it in my sleep. I talked about it to everyone, too. I would even ask a waiter whether he used Google when I ate at a restaurant.

In the meantime, whenever an employee suggested a new product idea, I always said, "Hold off! Let's wait until we meet our goals in search."

That made some employees unhappy. Chinese media also criticized us for not living up to the creative image of Google. Worst of all, some surveys indicated our market share to be dropping! A few employees felt disappointed and resigned.

Despite all the pressure from outside and within the company, I knew I was doing the right thing. Fortunately, it didn't take me too long to prove that. In October 2006, our four evaluation metrics in search exceeded our expectations. More significantly, in April 2007, we found ourselves making more progress in accuracy than Google's other foreign language search teams. Our Chinese web page indexes doubled within a year (and later grew 10 folds in 2008). We were able to reach new content of a website only a few minutes after the update (and later that became a few seconds in 2009), while web spam was reduced by 97.5%.

In June 2007, we received our ICP license from the Chinese government. Then Google moved Chinese language servers to China. Finally, anyone using google.com in China would be instantly directed to google.cn (with an option to opt out).

As we heard more and more compliments on our search results, our market share kept climbing in 2007 and 2008. That was when I thought we were truly ready to develop new products.

# Self Management

Google is famous for its egalitarian culture that gives engineers all the space they need to maximize their creativity without interference from management. This is exactly opposite to the Chinese concept of hierarchy, with which Google China's engineers were raised. These engineers came in thinking they were supposed to just follow their managers' instructions. At first I didn't bother converting them to the headquarters' culture because we had to concentrate on improving search technology. But once our search was on track, I began telling them to lead the company instead of being led.

I told them that they would have the right to transfer to another department or choose another project after the completion of their first year. However, they didn't seem to believe they really had so much freedom of choice. At the end of Google China's first year, my chief of staff, Ning Tao, reported to me that none of the engineers requested any change.

"They must have certain concerns that keep them from expressing what they want," I said to Ning. "Please ask them one by one to find out why they don't dare to request a transfer."

Ning did approach them one by one, and then she heard all kinds of concerns from the engineers. "If I requested transferring to another department, what would my boss think?" "I'm not sure if I can still get promoted if I transfer to another department." "If I work for another department, will there be a change to my salary?"

To address these concerns, Ning and I held a meeting with all the engineers to explain the policy that was meant to let them do what they found most interesting in order to bring out their best potential. We assured them that a transfer would definitely not affect their salaries or promotions.

After the meeting, Ning helped seven engineers change their projects. All the other engineers saw them prosper in their new positions, and realized their fear of the former manager's "revenge" was truly non-existent at Google. At the end of Google China's second year, more engineers requested transfers. One of them said to me, "I never imagined I could publicly express my preference and choose my preferred project as an engineer. It feels so nice to be so much valued!"

Google China's engineers gradually learned to voice their opinions. One day in 2007, an engineer approached me and said, "Kai-Fu, the survey from the headquarters looked too boring, and the questions didn't really address our issues. Can I conduct an extended survey?"

I was thrilled to hear such an initiative. "Go ahead!" I said with utmost encouragement. "Take complete charge of it! Management will not participate in or interfere with the process of the survey. But I'd like to see the results so I'll know what everyone wants me to do."

The engineer did make very interesting survey questions. Below are two of the multiple-choice questions and the statistics of their answers:

*What makes you the happiest about being a Google engineer?*

| | |
|---|---|
| *Writing programs* | *61.9%* |
| *Discussing technology* | *65.5%* |
| *Writing papers* | *0.9%* |
| *Learning* | *42.5%* |
| Interviewing | 2.7% |
| Business trips to America | 23.9% |
| Sleep | 17.7% |
| Browsing the web | 13.3% |
| Chatting on line | 7.1% |
| Chatting in the office | 22.1% |
| Dining in Google's cafeteria | 21.2% |
| Working with cool guys and cute girls | 9.7% |
| Playing games | 15.9% |
| Sports activities | 24.8% |

What makes you suffer most as a Google engineer?

| | |
|---|---|
| *Being Called to the manager's office* | *1.8%* |
| *A too busy schedule* | *25.7%* |
| *Program review* | *27.4%* |
| *Writing programs* | *2.7%* |
| *Writing documents* | *32.7%* |
| *Technical discussions* | *2.7%* |
| *Disagreement with senior staff* | *12.4%* |
| *Disagreement with tech lead* | *10.6%* |
| *A good product not launched* | *24.8%* |
| *Ideas not turned into products* | *31.0%* |
| *Endless reviewing process* | *41.6%* |
| *Endless interviews* | *14.2%* |
| *No one to talk to* | *3.5%* |
| *No time to find the special someone* | *6.2%* |
| *Being misunderstood by former colleagues* | *3.5%* |
| *Being misunderstood by other departments* | *3.5%* |
| *Nothing to do* | *8.8%* |

I learned a lot from the unofficial survey. Then I took action to address the issues coming out of it. I built an internal forum for employees to express any opinions they might have, and I created more opportunities for departments to communicate with one another. Most importantly, I officially encouraged all the engineers to spend 20% of their work time on their own creative ideas like their counterparts in the headquarters.

I was happy to see in the survey that some of Google China's engineers were already practicing the 20%-time rule. That meant these Chinese engineers had learned to be autonomous. It was time to make sure all Google China's engineers know they had such license.

Most people may not know that many of Google's amazing products, such as Gmail and Google News, came from some engineers' 20% time. This proves that Google's genius engineers can shine even more brilliantly when they don't have to worry about whether their manager will approve a project they want to do, whether they can get funding for the project they are doing, or whether their project can turn into a profit making product. I wanted the same to happen for our engineers at Google China.

However, it didn't seem as easy for the Chinese engineers to come up with their own projects as their counterparts in the headquarters. It was easier for me to tell them to concentrate on search in 2006 than to make them spend 20% of their work time on their individual creations in 2007. Again, this was mostly because they had been raised in a hierarchy-oriented culture and were more used to following instructions than taking initiatives. Another reason was their fear of failure. They had been regarded as geniuses all their lives, so they didn't want to risk breaking their perfect records by exploring something new with an unknown prospect.

Even though some of them were already taking the 20% time when I made it official, most of them hadn't done it and were reluctant to start. Those with busy team projects especially preferred not taking time away from their assignments. To motivate them to seriously take the 20% time, I asked all the department managers to list "20% time performance" as part of every employee's evaluation.

To help Google China's engineers understand the significance of the 20% time, I invited Christophe Bisciglia, a pioneer in cloud computing and a senior engineer at Google's headquarters, to give them a speech. When Chistophe faced our group, he started by asking, "May I ask which ones of you are not engineers? Managers, would you please raise your hands?"

All the management staff, including me, raised our hands. Then Christophe said with a smile, "All the managers, please leave." After we walked out of the conference room, he said to all our engineers, "This is what 20% time truly means. Your manager has no right to participate!"

From then on, Google China's engineers became more and more enthusiastic about their 20%-time projects. Successes emerged one after another.

One of the successful examples was an interactive map of open railroads during heavy snow. Li Shuangfeng, an engineer who was unable to go home for Chinese New Year's because of snow storms, came up with the idea while having lunch with a few other engineers stuck in Beijing for Chinese New Year's for the same reason. Shuangfeng said, "I just checked on line and found messages about open and closed railroads here and there. Why don't we integrate all the information and put it all on a map?"

His colleagues liked the idea, which happened to be quite feasible with Google map products handy. They worked with Shuangfeng to make it happen by 5 p.m. that day, and the map was displayed on the home page of google.cn the next day. It immediately became a hit.

At the end of the day, the engineers excitedly said to me, "This product only took 24 hours to launch, and it attracted seven million users within a day!"

Another important product was created for the victims of the May 2008 earthquake. After the severe earthquake, we first made maps showing the latest situation of the disaster zone and indicating what part of the area needed what kinds of supplies. Then we realized those who had relatives living in the disaster zone were anxiously looking for them, so we developed "search for relatives." We worked day and night to collect data from hospitals and shelters of the disaster zone. Eventually, our database covered more than 40 hospitals and more than 40,000 victims of the disaster zone.

One of our engineers, Feng Zhengzhu, said he was afraid not everyone knew about this new search service, so he wrote a program to gather on-line announcements made by those looking for relatives in the disaster zone, and matched 9,000 unanswered ones with the data we had. When there were matches, our engineers would email the announcement writers about the status and location of their relatives. Our efforts were broadly appreciated. Even a Wall Street Journal correspondent heard about it and mentioned it in an earthquake-related article.

In 2008, Google China also launched SMS Search, SMS Holiday Greetings, Forum Search, a translation service, and many other products. In the meantime Google Music was being incubated as the next big thing.

When Alan Guo, Feng Hong and Bin Lin first brought up the idea of an MP3 music downloading service in 2007, my initial reaction was a big "No!" I believed in staying away from music downloading for its inevitable involvement with copyright issues.

Google with its motto "Don't be evil" definitely wouldn't let users download pirated music. But if we bought music and charged users for downloading, our users in China probably wouldn't go for it because they could easily download free pirated music elsewhere. It just seemed to me that there was no way we could win in a music service.

"Kai-Fu," Alan said. "I think we can negotiate with record companies. If we get authorizations from them, we can legally let our users download music for free."

"There are hundreds of record companies in China. You are going to talk to all of them?" I shook my head. "Not to mention two record companies are involved in a lawsuit with our headquarters. That definitely makes it more difficult. It's a mission impossible!"

Alan remained optimistic. He said, "Kai-Fu, we can find a partner who has authorizations from record companies. Would you please let me try?"

I didn't respond for a few seconds. Then I said, "You have a very good idea, and I shouldn't interfere because it's your 20% time project. I just hope you realize how challenging this is going to be. If you recognize all the difficulties but still want to go ahead, I'll support you!"

At this moment, Professor Reddy came to my mind. In my head I heard his voice, "I disagree with you, but I support you." I told myself I would do the same for my subordinates.

In my heart I always knew challenging the impossible actually meant a possibility of creating a miracle. I had done it myself. How could I not let them take their shot at it?

It took Alan, Feng and Bin 10 months to seek a partner until they found a music website with a Chinese name, *Ju-Jing*, which means "a giant whale" in Chinese. The company's English name is Top100.

Top100 was founded to advocate music copyright in China. One of its large shareholders was the most famous Chinese basketball player Yao Ming. By November 2007, Top100 had been authorized by three large record companies and more than 30 smaller record companies to let the website's members download copyrighted songs for free after paying their membership fee. However, there was too much free pirated music on line for Top100 to attract members. Google happened to be able to offer Top100 the kind of help it needed. The two companies' cooperation was a win-win decision.

Google persuaded Top100 to stop charging membership fees and let everyone download copyrighted songs for free while building a profit-making business model on advertising. From November 2007 to March 2009, Top100 and Google obtained authorizations from four major record companies and more than 140 smaller record companies in the world to let the search engine's users download 3.5 million copyrighted songs at no cost.

It was groundbreaking for Chinese Internet users to download free copyrighted songs and for the song writers to share the revenue from advertising. Nothing worked better than this to solve the music pirating problem in China.

Many Chinese called it a miracle. It was a miracle coming out of Google China's 20% time.

Numerous reporters have asked me if it's too big of an investment for Google to let engineers take one day out of every five working days away from company-assigned projects. They also asked about the financial return of such an investment. Then I told them the system was not about money, but to present a liberating working environment that would attract creative geniuses.

Google co-founder Sergey Brin once said, "Our company's creativity comes from our employees. If we ever go through a bottleneck, it must be because we haven't been able to hire the best and brightest engineers fast enough. So, we must do everything to keep our employees."

The 20% time is definitely one of Google's most attractive features to many talented engineers. I am proud of having transplanted it into Google China successfully, and happy to see it continue to grow.

## Sticking to Google's Motto

In February 2009, around Chinese New Year's, Google China's annual kick-off party took place at Shangri-La hotel in Beijing. The entire staff dressed up for the occasion as Americans would for a Christmas party.

I walked onto a large stage. Facing hundreds of staff members, I said, "At our last kick-off party, when we celebrated our market share growth of 2007, some people said we owed our success to the disappearance of a few small search engines. They predicted we wouldn't be able to take market share away from our largest competitor in 2008. But we did it! This proves we were right about focusing on search. In the future, we need to pay more attention to mobile search, because a new era of telecommunication has begun."

The staff applauded enthusiastically. They knew as well as I did that our success didn't come easily. So many other international web service companies had failed to compete with local businesses in China. Amazon, eBay, MySpace and Yahoo! were all famous examples.

Google had gone through a tough time in China as well. But Google had more patience than other companies to stick it out. Then-CEO Eric Schmidt once said, "China has five thousand years of history, so we'll take five thousand years of patience to make a success in China."

None of us could see five thousand more years, but we already made a difference in three years. I believe this had a lot to do with Google's motto, "Don't be evil."

In "Letter from the Founders," a public letter addressed to potential shareholders, Google co-founder Larry Page elaborates on the significance of the motto:

Don't be evil. We believe strongly that in the long term, we will be better served—as shareholders and in all other ways—by a company that does good things for the world even if we forgo some short term gains. This is an important aspect of our culture and is broadly shared within the company.

*Google users trust our systems to help them with important decisions: medical, financial and many others. Our search results are the best we know how to produce. They are unbiased and objective, and we do not accept payment for them or for inclusion or more frequent updating. We also display advertising, which we work hard to make relevant, and we label it clearly. This is similar to a newspaper, where the advertisements are clear and the articles are not influenced by the advertisers' payments. We believe it is important for everyone to have access to the best information and research, not only to the information people pay for you to see.*

I insisted on keeping the same principle at Google China. We separated advertisements from search results and only displayed those relevant to the search key words. While other Chinese search engines charged for placing an advertisement in a higher search position, we refused to do so. We chose to push money away for keeping our search results fair and objective.

"Stop asking when Google China will start making a profit," I kept telling staff members in our toughest days. "Forget eye-catching products! Forget advertising! Just remember the quality of our search. Put our users first. Someday they will give us the recognition we deserve"

That day finally came. Starting in 2007, our users drastically and continuously increased. We called 2007 our product year, in which we launched 24 new products and services. Most of these products were localized, and many were specifically created for Chinese users. Google's unique system allowed us to design our own products without asking the headquarters for approval, whereas other international companies in general only let local staff take charge of sales. The space Google's headquarters gave us was where our success originated.

While autonomously creating new products, we followed the headquarters in every move of search technology. As soon as then-CEO Eric Schmidt brought up the idea of "universal search," we began working on integrating all the relevant information into our search results. For instance, if a user typed "Citibank," the results would include not only the latest news about the bank but also the bank's stock projections and share prices.

Given this direction in search, we called 2008 our integration year, by the end of which a scandal about Internet advertising in China brought the public's attention to the fairness of ad placement at websites. More and more Chinese users appreciated Google's way of drawing a line between advertisements and search results.

Google China's market share grew from 16.1% in 2006 to 31.0% in 2009. Although we still didn't get China's largest market share in search, our Google Maps, Google Mobile Maps, Google Mobile Search and Google Translate all became number one in 2009.

At our 2009 kick-off party, our product manager, Faye Xu, gave a Powerpoint presentation, in which a cartoon character was using all kinds of Google products for dating. He played Google Music for his girlfriend, obtained directions from Google Mobile Maps on a date and shopped for romantic gifts on Google Mobile Search. In the end, he put a ring he had found on Google on her finger.

The last page of the Powerpoint displayed a line, "Pervasive through Focus."

Everyone laughed and clapped. Suddenly, my eyes watered. I held back the tears of joy and smiled.

## Good-bye, Google!

I completed a four-year term as president of Google China. In my third year I also took charge of Google's Korean team and the fast growing Southeast Asia market. But I soon realized that it was never my dream or forte to manage a large staff. What I truly enjoyed doing was pioneer work, to explore uncharted territories and create something out of nothing.

Since I shocked the media by announcing my resignation on Sept. 4, 2009, many reporters have asked me to comment on my four years with Google China. I generally would summarize my view into three points:

1. It takes a lot of commitment and patience to run an international company in China. Google has shown enough patience and a business model suitable for Chinese users. That's why Google China began to see profit in its fourth year.
2. Product development has to be quick in China for the market's ever-changing eco-system. While it's great that Google allows the China team to design all kinds of new products, most of the products have to go through the headquarters' evaluations before their releases. Although this is for quality assurance, it inevitably slows Google China down.
3. It is commendable that Google gives the local team more license than most other international companies. That was what enabled Google China to create unique products like Google Music and invest in eight Chinese companies.

As for Google China's next step, I believe it should be brand building. Basically Google China has better products than its competitors, but the general public in China may not know enough about Google.

Many Chinese still regard Google as a faraway American brand with English products as its specialty.

To promote Google China, I sent engineers to lecture at more than 300 universities in my last year with the company. I also accepted TV interviews, after which there was a noticeable increase of users, so I stressed the importance of traditional media in China in my last report to the headquarters.

I hope Google China will continue to popularize itself through these channels if Google decides to keep the branch.

My four years at Google China will always remain one of my most treasured memories. In a way it has become part of me. The youthful energy I have taken away from Google will keep me going through endless challenges of my future journey, with a serious mindset but a playful heart!

*One of the many fun moments at Google China*

## CHAPTER 12

# Teacher Kai-Fu

During my years at Microsoft and Google, I devoted virtually all of my free time to advising Chinese college students.

I gave speeches from university to university in China. Sometimes I had to continuously fly from city to city because of my tight schedule. In April 2009, I flew to three cities and gave speeches at six universities within eight days. The speeches were all about how to make the best of college life, plan steps beyond college, grow as an intellectual through lifelong learning, lead a life of integrity, and set a meaningful career goal.

Every time I spoke to Chinese college students, I saw in their passionate eyes their eagerness to learn from me. I knew they were going through a crucial stage of their lives. They were wondering how they could use their college education to realize their dreams in the future, and they found guidance in my speeches.

*One of my popular speeches*
*Courtesy of cqwb.com.cn*

While Chinese universities don't provide as much counseling as their American counterparts, I have stepped up to try filling the void. It all started in 2000.

When I recruited interns for Microsoft Research Asia in 2000, I began to have close contact with many Chinese college students. I still clearly remember how an in-depth conversation I had with one of them shocked me.

The student said, "Kai-Fu, I hope I'll become as successful as you are, in as high a position as yours. It must feel very satisfying and powerful to manage so many people. Would you please tell me how I can climb the corporate ladder right to get to a senior executive's position?"

I didn't expect a straight-A student of a prestigious university to have such a shallow interpretation of success. But his way of thinking reflected the standard measurement of success in Chinese society, which simply judges students by their grades and adults based on the amount of money or power they have.

Many other Chinese students share the same view as the would-be executive because that's what their parents have instilled in them. It's hard for me to blame those parents because China was poor for too long and they had very difficult lives when they were young. Finally, the 21$^{st}$ century began with drastic improvements on the living standards in China. It's natural of the Chinese parents to wish for their children what they never had, which is to get rich by seizing opportunities in the fast growing economy.

However, the desire for quick wealth often clouds people's judgment. That's why piracy and corruption are rampant in China. It basically requires a complete change of mindset to truly eradicate China's piracy and corruption problems, and it should start with the upcoming generation.

Chinese college students of the 21$^{st}$ century have much better resources and many more choices than their parents did. But while the improved living standards and more open political climate give them the freedom of choice, they don't necessarily have the wisdom and experience to choose what's really right for them.

The students feel that I have what it takes to help them. I know Chinese culture well enough to understand where they are coming from, and I know Western culture well enough to show them how to be a world citizen in this era of globalization. I am viewed as "successful"; therefore my words would be credible. I am among the few "successful" people who are also willing to spend time with them.

The more I think about it, the more I realize why it was my father's last wish for me to help make China a better place. He wisely saw the capability of accomplishing this mission in me.

As a Chinese American, I see making China grow intellectually as helping America as well. It is certainly in America's best interest to have the world's largest population as a friendly ally rather than a potential threat. A more enlightened China will have a more understanding, more stable, and more mutually beneficial relationship with the United States.

To benefit not only China but also America and eventually the whole world, I have kept advising Chinese college students since 2000, the year I published a public letter in Chinese titled, "My first Letter to Chinese Students---Let's Start with a Discussion of Honesty and Integrity." By now I have published seven public letters addressing issues of Chinese college students. I have also been giving more than 300 speeches at Chinese universities every year.

## Planning for a Dream University

Every time after giving a speech at a university, I would have a chat with the university president and a few deans. They have all complained to me about the less than satisfactory quality of their faculty. They have told me that good students won't stay to teach after their graduation because the professors' salaries are low. The low salaries result in the lack of excellent professors. The professors who are not knowledgeable enough in their specialized fields make their students unhappy and lower the social status of professors in general. That in turn makes it difficult to raise professors' salaries. All these situations have formed a vicious cycle.

While complaining about the lack of highly qualified faculty, these presidents and deans would show me around and rave about their new buildings. That has made me sigh every time after speaking with them. I wonder if China's university presidents will ever take pride in the renowned new professors they have hired instead of the polished new buildings on their campuses. When will every college graduate in China see teaching as a desirable profession instead of buying into that saying, "Those who can't do, teach"?

One day I felt compelled to write down all my observations of Chinese college education's problems and my suggestions for solutions. I heard then-vice-premier Li Lanqing (Li being his family name as China's official English translation of his name follows the Chinese tradition to put family name first) cared deeply about education, so I wrote a letter to him.

Soon Mr. Li had one of his vice ministers contact me. The vice minister invited me to have dinner with him and said, "Vice Premier absolutely agrees with your comments on our education. He has passed your letter to those in charge of education policies."

The vice minister later arranged for me to see the vice premier in his office. We had a profound discussion of China's college education. He asked me to help him find feasible ways of improving Chinese universities.

After that, I began to do research on post-secondary education. Even though my day job was busier than ever after I transferred to the Microsoft headquarters, I squeezed time to read everything I could find about higher education and took lots of notes.

I read *The Use of the University* and *The Gold and the Blue: A Personal Memoir of the University of California* by Clark Kerr, former president of the University of California. I also read *Universities in the Marketplace, The Role of the American University*, and other books on contemporary issues of college education.

I've learned from the readings that 70% of the world's most academically successful universities and eight of the world's top 10 universities are in America. Most of the best students from China, Japan, India and numerous other countries choose to study at prestigious American universities.

How have American universities become a magnet to students worldwide? Through my personal research, I've found a unique feature of American higher education, which is the success of private universities. The top 10 American universities in humanities are all private. Nine of 10 top business schools in America are private. Five of America's top 10 engineering schools are private. The top three programs in biotech all belong to private universities. The top three programs in computer science also belong to private universities.

Why are private universities so successful in America? I think the major reason is their being less restricted than their public counterparts. Private universities don't need the government's approval to create a new department, raise the professors' salaries or launch reforms. When a subject area becomes promising, a private university can establish a new department for it and offer high salaries to hire the most highly qualified professors to teach it. When private universities give professors raises, they don't have to worry about whether the government will call it unfair to other universities. The flexible systems of private universities operate like the market economy, encouraging the survival of the fittest and the prosperity of the best.

The competition among private universities also raises their bars. As private universities work toward outperforming one another, they are all constantly making improvements.

Each of the most renowned private universities seems to have an outstanding special feature. For instance, Harvard has America's top-quality humanities programs. Stanford encourages innovation and incubates Silicon Valley's most brilliant entrepreneurs. Carnegie Mellon applies IT to every domain. Cal Tech operates like a small innovative laboratory.

Thanks to private universities, America's higher education remains the most desirable in the world despite the recent decline of the country's K-12 education. The model of American private universities is certainly worth adopting worldwide.

While America always attempts to bring democracy to other countries, I think it'll work much better if those less democratic countries establish high-quality universities first. Only well informed citizens can form a true democracy. If a country merely adopts America's political system but hasn't educated its citizens about democracy, the citizens won't know how to use their rights to vote. Then the politicians winning the elections are probably those who have bribed or brainwashed voters. They will become de facto dictators. What's the use of having a democratic façade there?

I have repeated again and again to those Americans who demand democracy of China that it can't happen overnight. It doesn't do China any real good to push for a quick change of its political system. As an old Chinese saying points out, "If you cook the same herbs in a different kind of broth, their effect will remain unchanged."

The most effective way to bring about a more democratic China is to improve China's higher education. This is a slow route but promises steady progress, and it was a major reason why I once attempted to found a world-class university in China.

I was planning on serving some of China's best students first and then setting an example for other Chinese universities, which would eventually benefit all students.

I talked to all my friends about the idea. Most of them were supportive. But some cautioned me that it would be an extremely difficult task.

"I know," I responded. "But its value just consists in its difficulty."

I had a strategy to overcome the largest barrier, the scarcity of highly qualified professors in China. I planned to start with only Ph. D. programs. In this case we would only need a few distinguished professors to teach in the beginning. Later we would establish masters programs and retain our Ph. D. graduates to teach them. Then we would gradually open undergraduate departments depending on how soon we would be able to recruit eligible professors for them.

I wrote a longer-than-100-page proposal of the university, using all my spare time from busy work with Microsoft. In the concluding paragraph of the proposal, I wrote:

As a classical Chinese poem puts it, "The vast river runs east; the currents wash away the millennium's best and brightest." However, a world class university will forever remain. I hope this world class university in China will serve as a beacon that lights up the future journey of China's education.

I carried the proposal around to seek funding for the university. During the holiday season of 2003, I traveled from Seattle to numerous cities in China and Taiwan. I also went to Hong Kong. Unfortunately, I faced one rejection after another.

Just when I felt the most frustrated, a friend in Hong Kong asked me to stay there for a few more days. He said his big boss wanted to see me and talk about the university project.

Two days later, I met the billionaire over lunch. He had read my entire proposal. He said, "Your idea sounds fantastic! This is different from the common practice of philanthropy. You will not only help people but also change the future of China's education. I know what a shame it is that China with the world's largest population has never had a world class university. I went to Westerners-run Catholic schools here in my childhood and then went to college in the West just because of that. I thought about funding the founding of such a university long ago but just didn't know anyone who could run the university. If you are willing to do it, I can donate 1.6 billion yuans (about 200 million US dollars)."

I could hardly believe how generous he was! Although the amount would only pay for the university's first few years, it was an enormous number coming out of one single donor.

Deeply moved, I said, "Since you are going to donate so much, the university should be named after you. Then people will always remember it was your generosity that made the first world class university appear in China!"

To my surprise, the elderly billionaire said, "No, Kai-Fu, please don't name it after me. I know 1.6 billion yuans will run out in a few years. If the university needs funding again and I'm no longer in this world, you'll have to look for donors again. If you ask others to donate to a university named after me, they might not be willing. But if it doesn't have my name on it, you can say you will name it after the next largest donor and hopefully get the funding you need."

I was stunned. Looking at the billionaire in his 80s, I saw the most precise definition of "big-heartedness." I thought of an English saying, "The richest man is not the one who has the most, but the one who needs the least."

I also recalled my father's motto:

Only with magnanimity can you absorb the greatest virtue
*Simply without desire will you build the highest character*

The elderly billionaire epitomized the greatest virtue and highest character by showing rare magnanimity and no desire.

# The Establishment of *Wo Xue*

With 1.6 billion yuans, I began looking for the site of my dream university in Beijing, Shanghai and Shenzhen. The local governments of these three cities all showed support for my project and offered me assistance.

However, I was unable to secure a long-term source of funding after receiving the 1.6 billion yuans. That made it nearly impossible for me to recruit the world's most distinguished professors. Without world class professors, how were we going to found a world class Ph. D. program?

Another challenge was getting government support. My supporter Vice Premier Li Lanqing was already retired by this time. His successor and other high ranking officials did not have the same excitement and foresight. I had a very hard time getting them to modify some rules to accommodate this great work.

I tried extremely hard until the end of 2004. Unfortunately, there were too many insurmountable obstacles. I had to eventually drop my dream project.

I was deeply disappointed. But at the same time I still kept receiving emails from Chinese college students. The emails reminded me that I was still able to help Chinese college students without founding a university.

I told myself, "Don't give up all your dreams just because one of them didn't come true!"

With that reassuring thought in mind, I expanded the Chinese language website I had created in 2003 to receive emails from Chinese college students and renamed it *Wo Xue*, which means "I am learning" in Chinese. The website address, www.5xue.com, is very easy for Chinese people to remember not only because it's short but also because number five in Chinese, *wu,* sounds very close to *wo,* which is "I" in Chinese.

The website currently has more than 500,000 registered visitors. I answer their questions about school, work, job hunting, studying abroad, personal growth, career planning, and even relationships. Sometimes their parents and professors communicate with me as well.

Since I have a day job, I must answer *Wo Xue* emails in my spare time. I get up very early every morning to check *Wo Xue* emails before going to work. I get on www.5xue.com at every airport where I'm waiting for a flight on a business trip or speech tour. I squeeze every possible moment to make sure I reply to every single email at the website.

I don't always have a perfect answer for each question I receive. But I try my best to provide my advice, or to at least point the students in the right direction to seek more specialized counseling. Most important of all, I make them feel they are not alone when feeling confused. There is someone they can feel free to talk to any time.

The students writing to me all call me "Teacher Kai-Fu" in their emails. In Chinese culture, "teacher" is a highly esteemed title, equivalent to "professor" or "Dr." in English. Chinese students usually put the word "teacher" together with a family name when addressing one of their teachers. But instead of "Teacher Lee", they call me "Teacher Kai-Fu." That means they see me not only as a teacher but also as a friend. I love that!

I recall in my college years, a professor asked us to envision our future epitaphs. He said, "An epitaph sums up a person's life. Knowing what you hope will be written in your epitaph will help you set your goals."

Since then I have often wondered what my epitaph will be. I know for space limitations on the tomb stone, the epitaph can only be a few lines. That means only a person's most important contribution to the world will be written in it. What will be counted as my most important contribution to the world?

Once I thought my epitaph would be:

**A scientist and entrepreneur**
*Who has worked for distinguished high tech companies*
*Transforming sophisticated technologies*
*Into user-friendly products that benefit everyone*

I still want those lines to be in my epitaph. But I feel I can contribute even more to China's education than to high tech for the rest of my life. Given the length limit of the epitaph, I may have to leave the lines about my achievement in high tech out and just take the following:

**A dedicated educator**
*Who has helped countless Chinese college students*
*Through writing, emailing and public speaking,*
*Whom every student affectionately called, "Teacher Kai-Fu"*

## Encouraging Everyone to Make a Difference

In my busy life, when I occasionally have a little free time to watch a movie, I often choose not to look for a new film, but to re-watch my all-time favorite, "Dead Poets Society."
A 1989 film starring Robin Williams and directed by Peter Weir, "Dead Poets Society" is set in 1959 at Welton Academy, an aristocratic boys' prep school in Vermont. The movie tells the story of an English teacher who inspires his students to change their lives of conformity through his teaching of literature.
The new English teacher, John Keating (Robin Williams), tells the students that they may call him "O Captain! My Captain!" (the title of a Walt Whitman poem) instead of Mr. Keating. For his first lesson, he whistles the *1812 Overture* and takes the students out of the classroom to focus on the idea of *carpe diem*, a Latin phrase that means "seize the day." In a later class, Keating has a student read the introduction to their poetry textbook, which describes how to rate the quality of poetry. Then Keating says such mathematical criticism is ridiculous, and he instructs the students to rip out the essay. He

also suggests the students try standing on their desks as a reminder to look at the world in a different way.

I've made myself a Chinese Keating since 2000. Just like Keating, I tell students not to follow others but to develop their own creativity and potential. I've published a book in Chinese, *Be Your Personal Best*, to advocate the idea to Chinese students.

Countless Chinese students have asked for my advice. Since so many of them trust me so much, I keep thinking, "What more can I do? What's the best way I can help them?"

While suggesting them find their inner voice and giving them recommendations on career planning, I've realized I don't necessarily know enough about every career path they consider taking. The career paths I have the most thorough knowledge of are those I've taken. Given my experiences as a computer scientist and senior executive, the most helpful advice I can provide is all related to high tech and business operations.

I've been fortunate to experience Apple and Microsoft in the era of personal computers and then Google in the era of the Internet. I've learned a great deal from these three world class companies and from world class entrepreneurs such as Steve Jobs, Bill Gates and Eric Schmidt. I regard my professional experiences as my most valuable asset, which I would love to pass on to the upcoming generation.

I think China's generation-Y business founders need my guidance more than their American counterparts because the channels for start-ups to take in China are less mature. I've confirmed this belief through communication with a few would-be Chinese entrepreneurs. One of them was a college students who had a severe disease. He travelled all the way to Beijing and asked to see me. He said he wouldn't start treatment until after talking to me about his business idea. When I did hear him out, I realized he knew nothing about starting up a company. He definitely needed a mentor!

Another one was my former subordinate at Microsoft. After I gave him some pointers on how to secure funding, he won the first place at an angel fund competition and received a decent amount of investment for his start-up. However, the products he attempted to make, despite being innovative, didn't really meet the needs of general consumers. I advised him to change directions, but he refused to listen. Then I couldn't help but watch his company fall into a crisis. In this case, he certainly needed a mentor who could help him choose the right project and make his innovation useful.

I also met a Chinese start-up founder with a degree from a prestigious American university. He persuaded me to let his productive new company become one of Google's business partners. We had pleasant cooperation at first. But two years later, a lawyer taught him to take advantage of a legal loophole and charge Google twice for the same service. That made me sigh deeply. He absolutely needed a mentor who would tell him any company aiming at long-term prosperity must keep honesty and integrity as core values!

These three young men will probably fail because they didn't have a suitable mentor by their side. Thinking about them motivated me to found a mentoring platform for potential entrepreneurs. That was how the idea of my new company, Innovation Works, began.

Since September 2009, I've been guiding young people with creative business ideas to work towards their dreams. I often remind myself, "If we are extremely lucky, we may create a world famous brand. If we are very lucky, we may establish a leading high tech company in China. But even if we are not so lucky, we can still spin off some valuable companies, make some useful products, and educate the next generation of entrepreneurs. All these results will make my life meaningful."

*Innovation Works serves as an incubator for creative new ventures*

When thinking about my mentor's role, I always feel as if being Mr. Keating standing on a desk in the movie "Dead Poets Society." The world indeed looks different from this point of view.

Like Mr. Keating in "Dead Poets Society," I love to recite a passage from *Walden* about the meaning of life, "I went to the woods because I wished to live deliberately, to front only the essential facts of life, and see if I could learn what it had to teach, and not, when I came to die, discover that I had not lived..."

"To live deliberately," I once kept urging myself to make a difference in the world. Then I began to encourage others to do the same. Enabling the dreams of my proteges now brings me more thrills than ever. Who knows if in the next five years the most innovative company in e-commerce, mobile Internet or cloud computing will come out of Innovation Works? Picturing this prospect motivates me to *carpe diem* every day, every moment.

I no longer focus on making a difference in the world purely through my own inventions. Now I also inspire others to invent, initiate and innovate. I say to everyone that I know has potential, "You can make a world of difference!"

# Credits

*To write this autobiography in Chinese and English, Dr. Kai-Fu Lee had assistance from a Chinese reporter and a Chinese American writer.*

## Co-author of the Chinese Autobiography

**Haitao Fan** began her career as a journalist at age 16, working as a student reporter for high school publications in Beijing. She went to Southwest Law and Political Science University, where she received a Bachelor of Arts degree and a law degree. She is currently a business news reporter with the Beijing Youth Daily, a renowned newspaper in China.

## Translator and cover designer of the Autobiography

**Crystal Tai** was born in Taiwan and immigrated to California with her family as a teenager. She holds a Master's degree in Education Policy, Organization and Leadership Studies with coursework in Journalism from Stanford University. She has been a reporter for the Silicon Valley Community Newspapers, which is affiliated with the San Jose Mercury News. Currently, she is a regular contributor to Patch.com, an AOL-owned, on-line news service. She has published three e-books in English, ***Secrets of Asian Women, Real Savings for Your Dream Wedding and A Poetic Portal to Chinese Culture***, with book covers all designed by herself. She also takes translation projects on various topics as a freelancer through her own website: www.crystaltai.com.